Experimental Techniques in High-Energy Nuclear and Particle Physics

Experimental Techniques in High-Energy Nuclear and Particle Physics

Contributors

Brian Robson et al.

AURIS
Reference

www.aurisreference.com

Experimental Techniques in High-Energy Nuclear and Particle Physics

Contributors: Brian Robson et al.

Published by Auris Reference Limited
www.aurisreference.com

United Kingdom

Experimental Techniques in High-Energy Nuclear and Particle Physics

ISBN: 978-1-78154-888-2

British Library Cataloguing in Publication Data
A CIP record for this book is available from the British Library

Printed in the United Kingdom

Exclusively distributed by CBS Publishers & Distributors Pvt. Ltd.

Sales & Distribution Rights only for India, Pakistan, Bangladesh, Sri Lanka, Nepal and Bhutan.This book is not to be sold outside these territories.

Contents

List of Abbreviations

CKM	Cabibbo-Kobayashi-Maskawa
CGM	Composite Generation Model
GM	Generation Model
SM	Standard Model
GWS	Weinberg and Salam
SEE	Single Event Effects
TID	Total ionization dose
MSE	Mean Square Error
QGP	Quark-Gluon Plasma
BBN	Big-Bang Nucleosynthesis
CCD	Charge-Coupled Device
CMB	Cosmic Microwave Background
LHC	Large Hadron Collider
MACHOs	Massive Astrophysical Compact Halo Objects
MCMCs	Monte Carlo Markov Chains
WIMPs	Weakly Interacting Massive Particles
APD	Avalanche Photodiodes
QED	Quantum Electrodynamics
ASGC	Academia Sinica Grid Centre
CDF	Collider Detector at Fermi-lab
DOE	Department of Energy
EVO	Enabling Virtual Organization
FBSNG	Farm Batch System Next Generation
GPU	Graphic Processing Unit
HEP	High-Energy Physics
KISTI	Korea Institute of Science and Technology Information
QCD	Quantum Chromo Dynamics
SIP	Session Initiation Protocol
LHC	Large Hadron Collider
VRVS	Virtual Room Videoconferencing System
WMS	Workload Management System
WLCG	Worldwide Large Hadron Collider Computing Grid
CERN	European Laboratory for Particle Physics
FDNPS	Fukushima Daiichi Nuclear Power Station
LEP	Large Electron Positron
NLL	Next-to-Leading-Logarithmic
SCET	Soft Collinear Effective Theory

List of Contributors

Brian Robson
Department of Theoretical Physics, Research School of Physics and Engineering, the Australian National University, Canberra Australia

Omid Zeynali
Islamic Azad University, Dashtestan Branch, Borazjan, Iran

Daryoush Masti
Islamic Azad University, Dashtestan Branch, Borazjan, Iran

Maryam Nezafat
Islamic Azad University, Dashtestan Branch, Borazjan, Iran

Alireza Mallahzadeh
Islamic Azad University, Dashtestan Branch, Borazjan, Iran

Mahmoud Y. El-Bakry
Department of Physics, Faculty of Sciences, Ain Shams University, Cairo, Egypt

El-Sayed A. El-Dahshan
Egyptian E-Learning University, Giza, Egypt
Department of Physics, Faculty of Education, Ain Shams University, Cairo, Egypt

Amr Radi
Department of Physics, Faculty of Education, Ain Shams University, Cairo, Egypt
The British University in Egypt (BUE), Cairo, Egypt

Mohamed Tantawy
Department of Physics, Faculty of Sciences, Ain Shams University, Cairo, Egypt

Moaaz A. Moussa
Department of Physics, Faculty of Sciences, Ain Shams University, Cairo, Egypt
Buraydah Colleges, East Qassim University, Buraydah, KSA

M. I. Haque
Department of Kulliyat, AK Tibbiya College, Aligarh Muslim University, Aligarh, India

M. Tariq
Department of Physics, Aligarh Muslim University, Aligarh, India

Tahir Hussain
Department of Applied Physics, Aligarh Muslim University, Aligarh, India

Teresa Marrodán Undagoitia and Ludwig Rauch
Max-Planck-Institut für Kernphysik, Saupfercheckweg 1, D-69117 Heidelberg, Germany

Paola La Rocca
Museo Storico della Fisica e Centro Studi e Ricerche "E.Fermi"
Department of Physics and Astronomy, University of Catania Italy

Francesco Riggi
Department of Physics and Astronomy, University of Catania Italy

Kihyeon Cho
Korea Institute of Science and Technology Information Republic of Korea

Joseph John Bevelacqua
Bevelacqua Resources USA

Anna Kulesza
Institute for Theoretical Physics, WWU M"unster, M"unster, D-48149 Germany

Leszek Motyka
Institute of Physics, Jagellonian University, S. Lojasiewicza 11, Krak´ow, 30-348 Poland

Tomasz Stebel
Institute of Physics, Jagellonian University, S. Lojasiewicza 11, Krak´ow, 30-348 Poland

Vincent Theeuwes
Institute for Theoretical Physics, WWU M"unster, M"unster, D-48149 Germany

Preface

Experimental Techniques in High-Energy Nuclear and Particle Physics is a compilation of outstanding reviews of the ingenious methods developed for experimentation in modern nuclear and particle physics. This book provides a balanced view of the major tools and technical concepts currently in use, and elucidates the basic principles that underly the detection devices. The main purpose of first chapter is to present an alternative to the Standard Model (SM) of particle physics. The chapter presents an outline of the current formulation of the SM: the elementary particles and the fundamental interactions of the SM, and the basic problem inherent in the SM. Second chapter concerns on (TID), (DD) and (SEE) effects also high energy particles' effects on electronic properties of silicon. In third chapter, Genetic programming (GP) model has been used to discover a function that computes the rapidity distribution of created (total charged, positive and negative) pions for p—-Ar and p—-Xe collisions at 200 GeV/c and charged particles for p-pb collision at 5.02 TeV. In fourth chapter, an attempt is made to examine multifractality in multiparticle production in relativistic nuclear collisions; multifractality is investigated in 14.5 A GeV/c28Si-nucleus collisions. Fifth chapter summarizes the status of direct dark matter searches, focusing on the detector technologies used to directly detect a dark matter particle producing recoil energies in the keV energy scale. The phenomenological signal expectations, main background sources, statistical treatment of data and calibration strategies is discussed. The overall set of problems and solutions related to the use of Avalanche Photodiodes in the design, construction, test and operation of large electromagnetic calorimeters in nuclear and particle physics experiments, is described in sixth Chapter. In seventh chapter, we use the concept of e-Science to combine experiment, theory and computing in particle physics in order to achieve a more efficient research process. Particle physics applications are generally regarded as a driver for developing this global e-Science infrastructure. Eighth chapter focuses on muon colliders and their unique radiation characteristics, initial scoping calculations for tau colliders are presented. In ninth chapter, we discuss the impact of resummation on the numerical prediction for the associated Higgs boson production with top quarks at the LHC

Chapter 1

THE GENERATION MODEL OF PARTICLE PHYSICS

Brian Robson

Department of Theoretical Physics, Research School of Physics and Engineering, the Australian National University, Canberra Australia

INTRODUCTION

The main purpose of this chapter is to present an alternative to the Standard Model (SM) (Gottfried and Weisskopf, 1984) of particle physics. This alternative model, called the Generation Model (GM) (Robson, 2002; 2004; Evans and Robson, 2006), describes all the transition probabilities for interactions involving the six leptons and the six quarks, which form the elementary particles of the SM in terms of only three unified additive quantum numbers instead of the nine non-unified additive quantum numbers allotted to the leptons and quarks in the SM.

The chapter presents (Section 2) an outline of the current formulation of the SM: the elementary particles and the fundamental interactions of the SM, and the basic problem inherent in the SM. This is followed by (Section 3) a summary of the GM, highlighting the essential differences between the GM and the SM. Section 3 also introduces a more recent development of a composite GM in which both leptons and quarks have a substructure. This enhanced GM has been named the Composite Generation Model (CGM) (Robson, 2005; 2011a). In this chapter, for convenience, we shall refer to this enhanced GM as the CGM, whenever the substructure of leptons and quarks is important for the discussion. Section 4 focuses on several important consequences of the different paradigms provided by the GM. In particular: the origin of mass, the mass hierarchy of the leptons and quarks, the origin of gravity and the origin of apparent CP violation, are discussed. Finally, Section 5 provides a summary and discusses future prospects.

STANDARD MODEL OF PARTICLE PHYSICS

The Standard Model (SM) of particle physics (Gottfried and Weisskopf, 1984) was developed throughout the 20th century, although the current formulation was essentially finalized in the mid-1970s following the experimental confirmation of the existence of quarks (Bloom et al., 1969; Breidenbach et al., 1969). The SM has enjoyed considerable success in describing the interactions of leptons and the multitude of hadrons (baryons and mesons) with each other as well as the decay modes of the unstable leptons and hadrons. However the model is considered to be incomplete in the sense that it provides no understanding of several empirical observations such as: the existence of three families or generations of leptons and quarks, which apart from mass have similar properties; the mass hierarchy of the elementary particles, which form the basis of the SM; the nature of the gravitational interaction and the origin of CP violation.

In this section a summary of the current formulation of the SM is presented: the elementary particles and the fundamental interactions of the SM, and then the basic problem inherent in the SM.

Elementary particles of the SM In the SM the elementary particles that are the constituents of matter are assumed to be the six leptons: electron neutrino (v_e), electron (e^-), muon neutrino (v_μ), muon (μ^-), tau neutrino (v_τ), tau ($\tau-$) and the six quarks: up (u), down (d), charmed (c), strange (s), top (t) and bottom (b), together with their antiparticles. These twelve particles are all spin- 1 2 particles and fall naturally into three families or generations: (i) v_e, e^-, u, d ; (ii) $v_\mu, \mu^-, c, \bar{s}$; (iii) v_τ, τ^-, t, b . Each generation consists of two leptons with charges Q = 0 and Q = −1 and two quarks with charges Q = +2 3 and Q = −1 3 . The masses of the particles increase significantly with each generation with the possible exception of the neutrinos, whose very small masses have yet to be determined.

In the SM the leptons and quarks are allotted several additive quantum numbers: charge Q, lepton number L, muon lepton number L_μ, tau lepton number L_τ, baryon number A, strangeness S, charm C, bottomness B and topness T. These are given in Table 1. For each particle additive quantum number N, the corresponding antiparticle has the additive quantum number −N.

Table 1: SM additive quantum numbers for leptons and quarks

particle	Q	L	L_μ	L_τ	A	S	C	B	T
ν_e	0	1	0	0	0	0	0	0	0
e^-	-1	1	0	0	0	0	0	0	0
ν_μ	0	1	1	0	0	0	0	0	0
μ^-	-1	1	1	0	0	0	0	0	0
ν_τ	0	1	0	1	0	0	0	0	0
τ^-	-1	1	0	1	0	0	0	0	0
u	$+\frac{2}{3}$	0	0	0	$\frac{1}{3}$	0	0	0	0
d	$-\frac{1}{3}$	0	0	0	$\frac{1}{3}$	0	0	0	0
c	$+\frac{2}{3}$	0	0	0	$\frac{1}{3}$	0	1	0	0
s	$-\frac{1}{3}$	0	0	0	$\frac{1}{3}$	-1	0	0	0
t	$+\frac{2}{3}$	0	0	0	$\frac{1}{3}$	0	0	0	1
b	$-\frac{1}{3}$	0	0	0	$\frac{1}{3}$	0	0	-1	0

Table 1 demonstrates that, except for charge, leptons and quarks are allotted different kinds of additive quantum numbers so that this classification of the elementary particles in the SM is non-unified.

The additive quantum numbers Q and A are assumed to be conserved in strong, electromagnetic and weak interactions. The lepton numbers L, L_μ and L_τ are not involved in strong interactions but are strictly conserved in both electromagnetic and weak interactions. The remainder, S, C, B and T are strictly conserved only in strong and electromagnetic interactions but can undergo a change of one unit in weak interactions.

The quarks have an additional additive quantum number called "color charge", which can take three values so that in effect we have three kinds of each quark, u, d, etc. These are often called red, green and blue quarks. The antiquarks carry anticolors, which for simplicity are called antired, antigreen and antiblue. Each quark or antiquark carries a single unit of color or anticolor charge, respectively. The leptons do not carry a color charge and consequently do not participate in the strong interactions, which occur between particles carrying color charges.

Fundamental interactions of the SM

The SM recognizes four fundamental interactions in nature: strong, electromagnetic, weak and gravity. Since gravity plays no role in particle physics because it is so much weaker than the other three fundamental interactions, the SM does not attempt to explain gravity. In the SM the other three fundamental interactions are assumed to be associated with a local gauge field.

Strong interactions

The strong interactions, mediated by massless neutral spin-1 gluons between quarks carrying a color charge, are described by an SU(3) local gauge theory called quantum chromodynamics (QCD) (Halzen and Martin, 1984). There are eight independent kinds of gluons, each of which carries a combination of a color charge and an anticolor charge (e.g. red-antigreen). The strong interactions between color charges are such that in nature the quarks (antiquarks) are grouped into composites of either three quarks (antiquarks), called baryons (antibaryons), each having a different color (anticolor) charge or a quark-antiquark pair, called mesons, of opposite color charges. In the SU(3) color gauge theory each baryon, antibaryon or meson is colorless. However, these colorless particles, called hadrons, may interact strongly via residual strong interactions arising from their composition of colored quarks and/or antiquarks. On the other hand the colorless leptons are assumed to be structureless in the SM and consequently do not participate in strong interactions.

Electromagnetic interactions

The electromagnetic interactions, mediated by massless neutral spin-1 photons between electrically charged particles, are described by a U(1) local gauge theory called quantum electrodynamics (Halzen and Martin, 1984).

Weak Interactions

The weak interactions, mediated by the massive W^+, W^- and Z^0 vector bosons between all the elementary particles of the SM, fall into two classes: (i) charge-changing (CC) weak interactions involving the W^+ and W^- bosons and (ii) neutral weak interactions involving the Z^0 boson. The CC weak interactions, acting exclusively on left-handed particles and right-handed antiparticles, are described by an SU(2)L local gauge theory, where the subscript L refers to left-handed particles only (Halzen and Martin, 1984). On the other hand, the neutral weak interactions act on both left-handed and right-handed particles, similar to the electromagnetic interactions. In fact the SM assumes (Glashow, 1961) that both the Z 0 and the photon (γ) arise from a mixing of two bosons, W^0 and B^0, via an electroweak mixing angle θ_W

$$\gamma = B^0 \cos\theta_W + W^0 \sin\theta_W , \tag{1}$$

$$Z^0 = -B^0 \sin\theta_W + W^0 \cos\theta_W . \tag{2}$$

These are described by a $U(1) \times SU(2)_L$ local gauge theory, where the U(1) symmetry involves both left-handed and right-handed particles.

Experiment requires the masses of the weak gauge bosons, W and Z, to be heavy so that the weak interactions are very short-ranged. On the other hand, Glashow's proposal, based upon the concept of a non-Abelian SU(2) Yang-Mills gauge theory, requires the mediators of the weak interactions to be massless like the photon. This boson mass problem was resolved by Weinberg (1967) and Salam (1968), who independently employed the idea of spontaneous symmetry breaking involving the Higgs mechanism (Englert and Brout, 1964; Higgs, 1964). In this way the W and Z bosons acquire mass and the photon remains massless.

The above treatment of the electromagnetic and weak interactions in terms of a U(1) × SU(2)$_L$ local gauge theory has become known as the Glashow, Weinberg and Salam (GWS) model and forms one of the cornerstones of the SM. The model gives the relative masses of the W and Z bosons in terms of the electroweak mixing angle:

$$M_W = M_Z \cos \theta_W . \tag{3}$$

The Higgs mechanism was also able to cure the associated fermion mass problem (Aitchison and Hey, 1982): the finite masses of the leptons and quarks cause the Lagrangian describing the system to violate the SU(2)$_L$ gauge invariance. By coupling originally massless fermions to a scalar Higgs field, it is possible to produce the observed physical fermion masses without violating the gauge invariance. However, the GWS model requires the existence of a new massive spin zero boson, the Higgs boson, which to date remains to be detected. In addition, the fermion-Higgs coupling strength is dependent upon the mass of the fermion so that a new parameter is required for each fermion mass in the theory.

In 1971, t'Hooft (1971a,b) showed that the GWS model of the electroweak interactions was renormalizable and this self-consistency of the theory led to its general acceptance. In 1973, events corresponding to the predicted neutral currents mediated by the Z 0 boson were observed (Hasert et al., 1973; 1974), while bosons, with approximately the expected masses, were discovered in 1983 (Arnison et al., 1983; Banner et al., 1983), thereby confirming the GWS model.

Another important property of the CC weak interactions is their universality for both leptonic and hadronic processes. In the SM this property is taken into account differently for leptonic and hadronic processes. For leptonic CC weak interaction processes, each of the charged leptons is assumed to form a weak isospin doublet (i = 1/2) with its respective neutrino, i.e. $(\nu_e, e^-), (\nu_\mu, \mu^-), (\nu_\tau, \tau^-),$, with each doublet having the third component of weak isospin $i_3 = (+\frac{1}{2}, -\frac{1}{2})$. In addition each doublet is associated with a different lepton number so that there are no CC weak interaction transitions between generations.

Thus for leptonic processes, the concept of a universal CC weak interaction allows one to write (for simplicity we restrict the discussion to the first two generations only):

$$a(v_e, e^-; W^-) = a(v_\mu, \mu^-; W^-) = g_w .$$

(4)

Here a(α, β; W$^-$) represents the CC weak interaction transition amplitude involving the fermions α, β and the W$^-$ boson, and g$_w$ is the universal CC weak interaction transition amplitude. Lepton number conservation gives

$$a(v_e, \mu^-; W^-) = a(v_\mu, e^-; W^-) = 0 ,$$

(5)

so that there are no CC weak interaction transitions between generations in agreement with experiment.

Unlike the pure leptonic decays, which are determined by the conservation of the various lepton numbers, there is no quantum number in the SM which restricts quark (hadronic) CC weak interaction processes between generations. In the SM the quarks do not appear to form weak isospin doublets: the known decay processes of neutron β-decay and Λ^0 β-decay suggest that quarks mix between generations and that the "flavor" quantum numbers, S, C, B and T are not necessarily conserved in CC weak interaction processes.

In the SM neutron β-decay:

$$n^0 \rightarrow p^+ + e^- + \bar{v}_e ,$$

(6)

is interpreted as the sequential transition

$$d \rightarrow u + W^- , \quad W^- \rightarrow e^- + \bar{v}_e .$$

(7)

The overall coupling strength of the CC weak interactions involved in neutron β-decay was found to be slightly weaker (≈ 0.95) than that for muon decay:

$$\mu^- \rightarrow v_\mu + W^- , \quad W^- \rightarrow e^- + \bar{v}_e .$$

(8)

Similarly, Λ^0 β-decay:

$$\Lambda^0 \rightarrow p^+ + e^- + \bar{v}_e ,$$

(9)

is interpreted in the SM as the sequential transition

$$s \rightarrow u + W^- , \quad W^- \rightarrow e^- + \bar{v}_e .$$

(10)

In this case the overall coupling strength of the CC weak interactions was found to be significantly less (≈ 0.05) than that for muon decay.

In the SM the universality of the CC weak interaction for both leptonic and hadronic processes is restored by adopting the proposal of Cabibbo (1963) that in hadronic processes the CC weak interaction is shared between $\Delta S = 0$ and $\Delta S = 1$ transition amplitudes in the ratio of cos θc: sin θc. The Cabibbo angle θc has a value $\approx 13°$, which gives good agreement with experiment for the decay processes (7) and (10) relative to (8).

This "Cabibbo mixing" is an integral part of the SM. In the quark model it leads to a sharing of the CC weak interaction between quarks with different flavors (different generations) unlike the corresponding case of leptonic processes. Again, in order to simplify matters, the following discussion (and also throughout the chapter) will be restricted to the first two generations of the elementary particles of the SM, involving only the Cabibbo mixing, although the extension to three generations is straightforward (Kobayashi and Maskawa, 1973). In the latter case, the quark mixing parameters correspond to the so-called Cabibbo-Kobayashi-Maskawa (CKM) matrix elements, which indicate that inclusion of the third generation would have a minimal effect on the overall coupling strength of the CC weak interactions. Cabibbo mixing was incorporated into the quark model of hadrons by postulating that the so-called weak interaction eigenstate quarks, d ' and s ' , form CC weak interaction isospin doublets with the u and c quarks, respectively: (u, d ') and (c,s '). These weak eigenstate quarks are linear superpositions of the so-called mass eigenstate quarks (d and s):

$$d' = d \cos \theta_c + s \sin \theta_c$$

(11)

and

$$s' = -d \sin \theta_c + s \cos \theta_c .$$

(12)

The quarks d and s are the quarks which participate in the electromagnetic and the strong interactions with the full allotted strengths of electric charge and color charge, respectively. The quarks d ' and s ' are the quarks which interact with the u and c quarks, respectively, with the full strength of the CC weak interaction.

In terms of transition amplitudes, Eqs. (11) and (12) can be represented as

$$a(u, d'; W^-) = a(u, d; W^-) \cos \theta_c + a(u, s; W^-) \sin \theta_c = g_w$$

(13)

and

$$a(c, s'; W^-) = -a(c, d; W^-)\sin\theta_c + a(c, s; W^-)\cos\theta_c = g_w \ . \tag{14}$$

In addition one has the relations

$$a(u, s'; W^-) = -a(u, d; W^-)\sin\theta_c + a(u, s; W^-)\cos\theta_c = 0 \tag{15}$$

and

$$a(c, d'; W^-) = a(c, d; W^-)\cos\theta_c + a(c, s; W^-)\sin\theta_c = 0 \ . \tag{16}$$

Eqs. (13) and (14) indicate that it is the d ' and s ' quarks which interact with the u and c quarks, respectively, with the full strength gw. These equations for quarks correspond to Eq. (4) for leptons. Similarly, Eqs. (15) and (16) for quarks correspond to Eq. (5) for leptons. However, there is a fundamental difference between Eqs. (15) and (16) for quarks and Eq. (5) for leptons. The former equations do not yield zero amplitudes because there exists some quantum number (analagous to muon lepton number) which is required to be conserved. This lack of a selection rule indicates that the notion of weak isospin symmetry for the doublets (u, d ') and (c,s ') is dubious.

Eqs. (13) and (15) give

$$a(u, d; W^-) = g_w \cos\theta_c \ , \qquad a(u, s; W^-) = g_w \sin\theta_c \ . \tag{17}$$

Thus in the two generation approximation of the SM, transitions involving d $\rightarrow$ u + W$^-$ proceed with a strength proportional to $g_w^2 \cos^2\theta_c \approx 0.95 g_{w'}^2$, while transitions involving s $\rightarrow$ u + W$^-$ proceed with a strength proportional to $g_w^2 \sin^2\theta_c \approx 0.05 g_{w'}^2$, as required by experiment.

BASIC PROBLEM INHERENT IN SM

The basic problem with the SM is the classification of its elementary particles employing a diverse complicated scheme of additive quantum numbers (Table 1), some of which are not conserved in weak interaction processes; and at the same time failing to provide any physical basis for this scheme.

A good analogy of the SM situation is the Ptolemaic model of the universe, based upon a stationary Earth at the center surrounded by a rotating system of crystal spheres refined by the addition of epicycles (small circular orbits) to describe the peculiar movements of the planets around the Earth. While the Ptolemaic model yielded an excellent description, it is a complicated diverse scheme for predicting the movements of the Sun, Moon, planets and the stars around a stationary Earth and unfortunately provides no understanding of these complicated movements.

Progress in understanding the universe was only made when the Ptolemaic model was replaced by the Copernican-Keplerian model, in which the Earth moved like the other planets around the Sun, and Newton discovered his universal law of gravitation to describe the approximately elliptical planetary orbits. The next section describes a new model of particle physics, the Generation Model (GM), which addresses the problem within the SM, replacing it with a much simpler and unified classification scheme of leptons and quarks, and providing some understanding of phenomena, which the SM is unable to address.

GENERATION MODEL OF PARTICLE PHYSICS

The Generation Model (GM) of particle physics has been developed over the last decade. In the initial paper (Robson, 2002) a new classification of the elementary particles, the six leptons and the six quarks, of the SM was proposed. This classification was based upon the use of only three additive quantum numbers: charge (Q), particle number (p) and generation quantum number (g), rather than the nine additive quantum numbers (see Table 1) of the SM. Thus the new classification is both simpler and unified in that leptons and quarks are assigned the same kind of additive quantum numbers unlike those of the SM. It will be discussed in more detail in Subsection 3.1.

Another feature of the new classification scheme is that all three additive quantum numbers, Q, p and g, are required to be conserved in all leptonic and hadronic processes. In particular the generation quantum number g is strictly conserved in weak interactions unlike some of the quantum numbers, e.g. strangeness S, of the SM. This latter requirement led to a new treatment of quark mixing in hadronic processes (Robson, 2002; Evans and Robson, 2006), which will be discussed in Subsection 3.2.

The development of the GM classification scheme, which provides a unified description of leptons and quarks, indicated that leptons and quarks are intimately related and led to the development of composite versions of the GM, which we refer to as the Composite Generation Model (CGM) (Robson, 2005; 2011a). The CGM will be discussed in Subsection 3.3. Subsection 3.4 discusses the fundamental interactions of the GM.

Unified classification of leptons and quarks

Table 2 displays a set of three additive quantum numbers: charge (Q), particle number (p) and generation quantum number (g) for the unified classification of the leptons and quarks corresponding to the current CGM (Robson, 2011a). As for Table 1 the corresponding antiparticles have the opposite sign for each particle additive quantum number.

Table 2: CGM additive quantum numbers for leptons and quarks

particle	Q	p	g	particle	Q	p	g
ν_e	0	-1	0	u	$+\frac{2}{3}$	$\frac{1}{3}$	0
e^-	-1	-1	0	d	$-\frac{1}{3}$	$\frac{1}{3}$	0
ν_μ	0	-1	±1	c	$+\frac{2}{3}$	$\frac{1}{3}$	±1
μ^-	-1	-1	±1	s	$-\frac{1}{3}$	$\frac{1}{3}$	±1
ν_τ	0	-1	$0,\pm2$	t	$+\frac{2}{3}$	$\frac{1}{3}$	$0,\pm2$
τ^-	-1	-1	$0,\pm2$	b	$-\frac{1}{3}$	$\frac{1}{3}$	$0,\pm2$

Each generation of leptons and quarks has the same set of values for the additive quantum numbers Q and p. The generations are differentiated by the generation quantum number g, which in general can have multiple values. The latter possibilities arise from the composite nature of the leptons and quarks in the CGM.

The three conserved additive quantum numbers, Q, p and g are sufficient to describe all the observed transition amplitudes for both hadronic and leptonic processes, provided each "force" particle, mediating the various interactions, has $p = g = 0$.

Comparison of Tables 1 and 2 indicates that the two models, SM and CGM, have only one additive quantum number in common, namely electric charge Q, which serves the same role in both models and is conserved. The second additive quantum number of the CGM, particle number p, replaces both lepton number L and baryon number A of the SM. The third additive quantum number of the CGM, generation quantum number g, effectively replaces the remaining additive quantum numbers of the SM, L_μ, L_τ, S, C, B and T.

Table 2 shows that the CGM provides both a simpler and unified classification scheme for leptons and quarks. Furthermore, the generation quantum number g is conserved in the CGM unlike the additive quantum numbers, S, C, B and T of the SM. Conservation of g requires a new treatment of quark mixing in hadronic processes, which will be discussed in the next subsection.

Quark mixing in hadronic CC weak interaction processes in the GM

The GM differs from the SM in two fundamental ways, which are essential to preserve the universality of the CC weak interaction for both leptonic and hadronic processes. In the SM this was accomplished, initially by Cabibbo (1963) for the first two generations by the introduction of "Cabibbo quark mixing", and later by Kobayashi and Maskawa (1973), who generalized quark mixing involving the CKM matrix elements to the three generations.

Firstly, the GM postulates that the mass eigenstate quarks of the same generation, e.g. (u, d), form weak isospin doublets and couple with the full strength of the CC weak interaction, g_w, like the lepton doublets, e.g. (ve,e⁻). Unlike the SM, the GM requires that there is no coupling between mass eigenstate quarks from different generations. This latter requirement corresponds to the conservation of the generation quantum number g in the CC weak interaction processes.

Secondly, the GM postulates that hadrons are composed of weak eigenstate quarks such as d' and s' given by Eqs. (11) and (12) in the two generation approximation, rather than the corresponding mass eigenstate quarks, d and s, as in the SM.

To maintain lepton-quark universality for CC weak interaction processes in the two generation approximation, the GM postulates that

$$a(u,d;W^-) = a(c,s;W^-) = g_w$$

$$(18)$$

and generation quantum number conservation gives

$$a(u,s;W^-) = a(c,d;W^-) = 0.$$

$$(19)$$

Eqs. (18) and (19) are the analogues of Eqs. (4) and (5) for leptons. Thus the quark pairs (u, d) and (c,s) in the GM form weak isospin doublets, similar to the lepton pairs (v_e,e^-) and (v_μ, μ^-), thereby establishing a close lepton-quark parallelism with respect to weak isospin symmetry

To account for the reduced transition probabilities for neutron and Λ^0 β-decays, the GM postulates that the neutron and Λ^0 baryon are composed of weak eigenstate quarks, u, d ' and s' . Thus, neutron β-decay is to be interpreted as the sequential transition

$$d' \rightarrow u + W^-, \quad W^- \rightarrow e^- + \bar{v}_e.$$

$$(20)$$

The primary transition has the amplitude a (u, d'; W⁻) given by

$$a(u,d';W^-) = a(u,d;W^-)\cos\theta_c + a(u,s;W^-)\sin\theta_c = g_w\cos\theta_c,$$

$$(21)$$

where we have used Eqs. (18) and (19). This gives the same transition probability for neutron β-decay $(g_w^4\cos^2\theta_c)$ relative to muon decay (g_w^4) as the SM. Similarly, Λ^0 β-decay is to be interpreted as the sequential transition

$$s' \rightarrow u + W^-, \quad W^- \rightarrow e^- + \bar{v}_e.$$

$$(22)$$

In this case the primary transition has the amplitude a (u, s'; W⁻) given by

$$a(u, s'; W^-) = -a(u, d; W^-) \sin \theta_c + a(u, s; W^-) \cos \theta_c = -g_w \sin \theta_c .$$

(23)

Thus Λ^0 β-decay has the same transition probability $(g_w^4 \sin^2 \theta_c)$ relative to muon decay (g_w^4) as that given by the SM.

The GM differs from the SM in that it treats quark mixing differently from the method introduced by Cabibbo (1963) and employed in the SM. Essentially, in the GM, the quark mixing is placed in the quark states (wave functions) rather than in the CC weak interactions. This allows a unified and simpler classification of both leptons and quarks in terms of only three additive quantum numbers, Q, p and g, each of which is conserved in all interactions

COMPOSITE GENERATION MODEL

The unified classification scheme of the GM makes feasible a composite version of the GM (CGM) (Robson, 2005). This is not possible in terms of the non-unified classification scheme of the SM, involving different additive quantum numbers for leptons than for quarks and the non-conservation of some additive quantum numbers, such as strangeness, in the case of quarks. Here we shall present the current version (Robson, 2011a), which takes into account the mass hierarchy of the three generations of leptons and quarks. There is evidence that leptons and quarks, which constitute the elementary particles of the SM, are actually composites.

Firstly, the electric charges of the electron and proton are opposite in sign but are exactly equal in magnitude so that atoms with the same number of electrons and protons are neutral. Consequently, in a proton consisting of quarks, the electric charges of the quarks are intimately related to that of the electron: in fact, the up quark has charge Q = +2/3 and the down quark has charge Q = −1/3 , if the electron has electric charge Q = −1. These relations are readily comprehensible if leptons and quarks are composed of the same kinds of particles.

Secondly, the leptons and quarks may be grouped into three generations: (i) (v_e,e−, u, d), (ii) (v_μ, μ−, c, s) and (iii) (v_τ, τ −, t, b), with each generation containing particles which have similar properties. Corresponding to the electron, e −, the second and third generations include the muon, μ −, and the tau particle, τ −, respectively. Each generation contains a neutrino associated with the corresponding leptons: the electron neutrino, ve, the muon neutrino, vμ, and the tau neutrino, v_τ. In addition, each generation contains a quark with Q = +2/3 (the u, c and t quarks) and a quark with Q = −1/3 (the d, s and b quarks). Each pair of leptons, e.g. (v_e,e⁻), and each pair of quarks, e.g. (u, d), are connected by isospin symmetries, otherwise the grouping into the three families is according to increasing mass of the corresponding family members. The existence of three repeating patterns suggests strongly that the members of each generation are composites.

Thirdly, the GM, which provides a unified classification scheme for leptons and quarks, also indicates that these particles are intimately related. It has been demonstrated (Robson, 2004) that this unified classification scheme leads to a relation between strong isospin (I) and weak isospin (i) symmetries. In particular, their third components are related by an equation

$$i_3 = I_3 + \frac{1}{2}g ,$$

(24)

where g is the generation quantum number. In addition, electric charge is related to I_3, p, g and i_3 by the equations:

$$Q = I_3 + \frac{1}{2}(p+g) = i_3 + \frac{1}{2}p .$$

(25)

These relations are valid for both leptons and quarks and suggest that there exists an underlying flavor SU(3) symmetry. The simplest conjecture is that this new flavor symmetry is connected with the substructure of leptons and quarks, analogous to the flavor SU(3) symmetry underlying the quark structure of the lower mass hadrons in the Eightfold Way (Gell-Mann and Ne'eman, 1964).

The CGM description of the first generation is based upon the two-particle models of Harari (1979) and Shupe (1979), which are very similar and provide an economical and impressive description of the first generation of leptons and quarks. Both models treat leptons and quarks as composites of two kinds of spin-1/2 particles, which Harari named "rishons" from the Hebrew word for first or primary. This name has been adopted for the constituents of leptons and quarks. The CGM is constructed within the framework of the GM, i.e. the same kind of additive quantum numbers are assigned to the constituents of both leptons and quarks, as were previously allotted in the GM to leptons and quarks (see Table 2).

In the Harari-Shupe Model (HSM), two elementary spin-1/2 rishons and their corresponding antiparticles are employed to construct the leptons and quarks: (i) a T-rishon with Q = +1/3 and (ii) a V-rishon with Q = 0. Their antiparticles (denoted in the usual way by a bar over the defining particle symbol) are a $\bar{T}$-antirishon with Q = -1/3 and a $\bar{V}$-antirishon with Q = 0, respectively. Each spin-1/2 lepton and quark is composed of three rishons/antirishons. Table 3 shows the proposed structures of the first generation of leptons and quarks in the HSM.

Table 3: HSM of first generation of leptons and quarks

particle	structure	Q
e^+	TTT	$+1$
u	TTV, TVT, VTT	$+\frac{2}{3}$
$\bar{d}$	TVV, VTV, VVT	$+\frac{1}{3}$
ν_e	VVV	0
$\bar{\nu}_e$	$\bar{V}\bar{V}\bar{V}$	0
d	$\bar{T}\bar{V}\bar{V}, \bar{V}\bar{T}\bar{V}, \bar{V}\bar{V}\bar{T}$	$-\frac{1}{3}$
$\bar{u}$	$\bar{T}\bar{T}\bar{V}, \bar{T}\bar{V}\bar{T}, \bar{V}\bar{T}\bar{T}$	$-\frac{2}{3}$
e^-	$\bar{T}\bar{T}\bar{T}$	-1

It should be noted that no composite particle involves mixtures of rishons and antirishons, as emphasized by Shupe. Both Harari and Shupe noted that quarks contained mixtures of the two kinds of rishons, whereas leptons did not. They concluded that the concept of color related to the different internal arrangements of the rishons in a quark: initially the ordering TTV, TVT and VTT was associated with the three colors of the u-quark. However, at this stage, no underlying mechanism was suggested for color. Later, a dynamical basis was proposed by Harari and Seiberg (1981), who were led to consider color-type local gauged SU(3) symmetries, namely SU(3)C × SU(3)H, at the rishon level. They proposed a new super-strong color-type (hypercolor) interaction corresponding to the SU(3)H symmetry, mediated by massless hypergluons, which is responsible for binding rishons together to form hypercolorless leptons or quarks. This interaction was assumed to be analogous to the strong color interaction of the SM, mediated by massless gluons, which is responsible for binding quarks together to form baryons or mesons. However, in this dynamical rishon model, the color force corresponding to the SU(3)C symmetry is also retained, with the T-rishons and V-rishons carrying colors and anticolors. respectively, so that leptons are colorless but quarks are colored. Similar proposals were made by others (Casalbuoni and Gatto, 1980; Squires, 1980; 1981). In each of these proposals, both the color force and the new hypercolor interaction are assumed to exist independently of one another so that the original rishon model loses some of its economical description. Furthermore, the HSM does not provide a satisfactory understanding of the second and third generations of leptons and quarks.

Table 4: CGM additive quantum numbers for rishons

rishon	Q	p	g
T	$+\frac{1}{3}$	$+\frac{1}{3}$	0
V	0	$+\frac{1}{3}$	0
U	0	$+\frac{1}{3}$	-1

In order to overcome some of the deficiencies of the simple HSM, the two-rishon model was extended (Robson, 2005; 2011a), within the framework of the GM, in several ways.

Firstly, following the suggested existence of an SU(3) flavor symmetry underlying the substructure of leptons and quarks by Eq. (25), a third type of rishon, the U-rishon, is introduced. This U-rishon has $Q = 0$ but carries a non-zero generation quantum number, $g = -1$ (both the T-rishon and the V-rishon are assumed to have $g = 0$). Thus, the CGM treats leptons and quarks as composites of three kinds of spin-1/2 rishons, although the U-rishon is only involved in the second and third generations.

Secondly, in the CGM, each rishon is allotted both a particle number p and a generation quantum number g. Table 4 gives the three additive quantum numbers allotted to the three kinds of rishons. It should be noted that for each rishon additive quantum number N, the corresponding antirishon has the additive quantum number $-N$.

Historically, the term "particle" defines matter that is naturally occurring, especially electrons. In the CGM it is convenient to define a matter "particle" to have $p > 0$, with the antiparticle having $p < 0$. This definition of a matter particle leads to a modification of the HSM structures of the leptons and quarks which comprise the first generation. Essentially, the roles of the V-rishon and its antiparticle V^- are interchanged in the CGM compared with the HSM. Table 5 gives the CGM structures for the first generation of leptons and quarks. The particle number p is clearly given by 1 3 (number of rishons - number of antirishons). Thus the u-quark has $p = +1/3$, since it contains two T-rishons and one V^--antirishon. It should be noted that it is essential for the u-quark to contain a V^--antirishon ($p = -1/3$) rather than a V-rishon ($p = +1/3$) to obtain a value of $p = +1/3$, corresponding to baryon number $A = +1/3$ in the SM.

In the CGM, no significance is attached to the ordering of the T-rishons and the V^--antirishons (compare HSM) so that, e.g. the structures TTV^-, TV T^- and VTT^- for the u-quark are considered to be equivalent. The concept of color is treated differently in the CGM: it is assumed that all three rishons, T, V and U carry a color charge, red, green or blue, while their antiparticles carry an anticolor charge, antired, antigreen or antiblue. The CGM postulates a strong color-type interaction corresponding to a local gauged SU(3)C symmetry (analogous to QCD) and mediated by massless hypergluons, which is responsible for binding rishons and antirishons together to form colorless leptons and colored quarks. The proposed structures of the quarks requires the composite quarks to have a color charge so that the dominant residual interaction between quarks is essentially the same as that between rishons, and consequently the composite quarks behave very like the elementary quarks of the SM. In the CGM we retain the term "hypergluon" as the mediator of the strong color interaction, rather than the term "gluon" employed in the SM, because it is the rishons rather than the quarks, which carry an elementary color charge.

In the CGM each lepton of the first generation (Table 5) is assumed to be colorless, consisting of three rishons (or antirishons), each with a different color (or anticolor), analogous to the baryons (or antibaryons) of the SM. These leptons are built out of T- and V-rishons or their antiparticles $\bar{T}$ and $\bar{V}$, all of which have generation quantum number $g = 0$.

Table 5: CGM of first generation of leptons and quarks

particle	structure	Q	p	g
e^+	TTT	$+1$	$+1$	0
u	$TT\bar{V}$	$+\frac{2}{3}$	$+\frac{1}{3}$	0
$\bar{d}$	$T\bar{V}\bar{V}$	$+\frac{1}{3}$	$-\frac{1}{3}$	0
ν_e	$\bar{V}\bar{V}\bar{V}$	0	-1	0
$\bar{\nu}_e$	VVV	0	$+1$	0
d	$\bar{T}VV$	$-\frac{1}{3}$	$+\frac{1}{3}$	0
$\bar{u}$	$\bar{T}\bar{T}V$	$-\frac{2}{3}$	$-\frac{1}{3}$	0
e^-	$\bar{T}\bar{T}\bar{T}$	-1	-1	0

It is envisaged that each lepton of the first generation exists in an antisymmetric three-particle color state, which physically assumes a quantum mechanical triangular distribution of the three differently colored identical rishons (or antirishons), since each of the three color interactions between pairs of rishons (or antirishons) is expected to be strongly attractive (Halzen and Martin, 1984).

In the CGM, it is assumed that each quark of the first generation is a composite of a colored rishon and a colorless rishon-antirishon pair, $(T\bar{V})$ or $(V\bar{T})$, so that the quarks carry a color charge. Similarly, the antiquarks are a composite of an anticolored antirishon and a colorless rishon-antirishon pair, so that the antiquarks carry an anticolor charge.

In order to preserve the universality of the CC weak interaction processes involving first generation quarks, e.g. the transition d → u + W−, it is assumed that the first generation quarks have the general color structures:

up quark : $\quad T_C(T_{C'}\bar{V}_{C'})$, $\quad$ down quark : $\quad V_C(V_{C'}\bar{T}_{C'})$, $\quad$ with $C' \neq C$. $\qquad$ (26)

Thus a red u-quark and a red d-quark have the general color structures:

$$u_r = T_r(T_g\bar{V}_{\bar{g}} + T_b\bar{V}_{\bar{b}})/\sqrt{2} \, ,$$
$$\text{(27)}$$

and

$$d_r = V_r(V_g\bar{T}_{\bar{g}} + V_b\bar{T}_{\bar{b}})/\sqrt{2} \, ,$$
$$\text{(28)}$$

respectively. For d$_r$ → u$_r$ + W−, conserving color, one has the two transitions:

$$V_rV_g\bar{T}_{\bar{g}} \rightarrow T_rT_b\bar{V}_{\bar{b}} + V_rV_gV_b\bar{T}_{\bar{r}}\bar{T}_{\bar{g}}\bar{T}_{\bar{b}}$$
$$\text{(29)}$$

and

$$V_rV_b\bar{T}_{\bar{b}} \rightarrow T_rT_g\bar{V}_{\bar{g}} + V_rV_gV_b\bar{T}_{\bar{r}}\bar{T}_{\bar{g}}\bar{T}_{\bar{b}} \, ,$$
$$\text{(30)}$$

which take place with equal probabilities. In these transitions, the W− boson is assumed to be a three $\bar{T}$-antirishon and a three V-rishon colorless composite particle with additive quantum numbers Q = −1, p = g = 0. The corresponding W$^+$ boson has the structure $[T_rT_gT_b\bar{V}_{\bar{r}}\bar{V}_{\bar{g}}\bar{V}_{\bar{b}}]$, consisting of a colorless set of three T-rishons and a colorless set of three $\bar{V}$-antirishons with additive quantum numbers Q = +1, p = g = 0 (Robson, 2005).

Table 6: CGM of second generation of leptons and quarks

particle	structure	Q	p	g
μ^+	$TTT\Pi$	$+1$	$+1$	± 1
c	$TT\bar{V}\Pi$	$+\frac{2}{3}$	$+\frac{1}{3}$	± 1
$\bar{s}$	$T\bar{V}\bar{V}\Pi$	$+\frac{1}{3}$	$-\frac{1}{3}$	± 1
ν_μ	$\bar{V}\bar{V}\bar{V}\Pi$	0	-1	± 1
$\bar{\nu}_\mu$	$VVV\Pi$	0	$+1$	± 1
s	$\bar{T}VV\Pi$	$-\frac{1}{3}$	$+\frac{1}{3}$	± 1
$\bar{c}$	$\bar{T}\bar{T}V\Pi$	$-\frac{2}{3}$	$-\frac{1}{3}$	± 1
μ^-	$\bar{T}\bar{T}\bar{T}\Pi$	-1	-1	± 1

The rishon structures of the second generation particles are the same as the corresponding particles of the first generation plus the addition of a colorless rishon-antirishon pair, Π, where

$$\Pi = [(\bar{U}V) + (\bar{V}U)]/\sqrt{2},$$

$$(31)$$

which is a quantum mechanical mixture of (UV^-) and (VU^-), which have $Q = p = 0$ but $g = \pm 1$, respectively. In this way, the pattern for the first generation is repeated for the second generation. Table 6 gives the CGM structures for the second generation of leptons and quarks.

It should be noted that for any given transition the generation quantum number is required to be conserved, although each particle of the second generation has two possible values of g. For example, the decay

$$\mu^- \rightarrow \nu_\mu + W^-,$$

$$(32)$$

at the rishon level may be written

$$\bar{T}\bar{T}\bar{T}\Pi \rightarrow \bar{V}\bar{V}\bar{V}\Pi + \bar{T}\bar{T}\bar{T}VVV,$$

$$(33)$$

which proceeds via the two transitions:

$$\bar{T}\bar{T}\bar{T}(\bar{U}V) \rightarrow \bar{V}\bar{V}\bar{V}(\bar{U}V) + \bar{T}\bar{T}\bar{T}VVV$$

$$(34)$$

and

$$\bar{T}\bar{T}\bar{T}(\bar{V}U) \rightarrow \bar{V}\bar{V}\bar{V}(\bar{V}U) + \bar{T}\bar{T}\bar{T}VVV,$$

$$(35)$$

which take place with equal probabilities. In each case, the additional colorless rishon-antirishon pair, (UV^-) or (VU^-), essentially acts as a spectator during the CC weak interaction process. T

he rishon structures of the third generation particles are the same as the corresponding particles of the first generation plus the addition of two rishon-antirishon pairs, which are a quantum mechanical mixture of (UV^-) and (VU^-) and, as for the second generation, are assumed to be colorless and have $Q = p = 0$ but $g = \pm 1$. In this way the pattern of the first and second generation is continued for the third generation. Table 7 gives the CGM structures for the third generation of leptons and quarks.

Table 7: CGM of third generation of leptons and quarks

particle	structure	Q	p	g
τ^+	$TTTIII$	$+1$	$+1$	$0, \pm 2$
t	$TT\bar{V}III$	$+\frac{2}{3}$	$+\frac{1}{3}$	$0, \pm 2$
$\bar{b}$	$T\bar{V}\bar{V}III$	$+\frac{1}{3}$	$-\frac{1}{3}$	$0, \pm 2$
ν_τ	$\bar{V}\bar{V}\bar{V}III$	0	-1	$0, \pm 2$
$\bar{\nu}_\tau$	VVVIII	0	$+1$	$0, \pm 2$
b	$\bar{T}VVIII$	$-\frac{1}{3}$	$+\frac{1}{3}$	$0, \pm 2$
$\bar{t}$	$\bar{T}\bar{T}VIII$	$-\frac{2}{3}$	$-\frac{1}{3}$	$0, \pm 2$
τ^-	$\bar{T}\bar{T}\bar{T}III$	-1	-1	$0, \pm 2$

The rishon structure of the τ^+ particle is and each particle of the third generation is a similar quantum mechanical mixture of $g = 0, \pm 2$ components.

$$TTT\Pi\Pi = TTT[(\bar{U}V)(\bar{U}V) + (\bar{U}V)(\bar{V}U) + (\bar{V}U)(\bar{U}V) + (\bar{V}U)(\bar{V}U)]/2 \qquad (36)$$

The color structures of both second and third generation leptons and quarks have been chosen so that the CC weak interactions are universal. In each case, the additional colorless rishon-antirishon pairs, (UV^-) and/or (VU^-), essentially act as spectators during any CC weak interaction process. Again it should be noted that for any given transition the generation quantum number is required to be conserved, although each particle of the third generation now has three possible values of g. Furthermore, in the CGM the three independent additive quantum numbers, charge Q, particle number p and generation quantum number g, which are conserved in all interactions, correspond to the conservation of each of the three kinds of rishons (Robson, 2005):

$$n(T) - n(\bar{T}) = 3Q, \qquad (37)$$

$$n(\bar{U}) - n(U) = g, \qquad (38)$$

$$n(T) + n(V) + n(U) - n(\bar{T}) - n(\bar{V}) - n(\bar{U}) = 3p, \qquad (39)$$

where n(R) and n(R$^-$) are the numbers of rishons and antirishons, respectively. Thus, the conservation of g in weak interactions is a consequence of the conservation of the three kinds of rishons (T, V and U), which also prohibits transitions between the third generation and the first generation via weak interactions even for $g = 0$ components of third generation particles.

Fundamental Interactions of the GM

The GM recognizes only two fundamental interactions in nature: (i) the usual electromagnetic interaction and (ii) a strong color-type interaction, mediated by massless hypergluons, acting between color charged rishons and/or antirishons.

The only essential difference between the strong color interactions of the GM and the SM is that the former acts between color charged rishons and/or antirishons while the latter acts between color charged elementary quarks and/or antiquarks. For historical reasons we use the term "hypergluons" for the mediators of the strong color interactions at the rishon level, rather than the term "gluons" as employed in the SM, although the effective color interaction between composite quarks and/or composite antiquarks is very similar to that between the elementary quarks and/or elementary antiquarks of the SM.

In the GM both gravity and the weak interactions are considered to be residual interactions of the strong color interactions. Gravity will be discussed in some detail in Subsection 4.3. In the GM the weak interactions are assumed to be mediated by composite massive vector bosons, consisting of colorless sets of three rishons and three antirishons as discussed in the previous subsection, so that they are not elementary particles, associated with a U(1) × SU(2)L local gauge theory as in the SM. The weak interactions are simply residual interactions of the CGM strong color force, which binds rishons and antirishons together, analogous to the strong nuclear interactions, mediated by massive mesons, being residual interactions of the strong color force of the SM, which binds quarks and antiquarks together. Since the weak interactions are not considered to be fundamental interactions arising from a local gauge theory, there is no requirement for the existence of a Higgs field to generate the boson masses within the framework of the GM (Robson, 2008).

CONSEQUENCES

In this section it will be shown that new paradigms arising from the GM provide some understanding concerning: (i) the origin of mass; (ii) the mass hierarchy of leptons and quarks; (iii) the origin of gravity and (iv) the origin of "apparent" CP violation in the $K^0 - \overline{K}^0$ system.

Origin of Mass

Einstein (1905) concluded that the mass of a body m is a measure of its energy content E and is given by

$$m = E/c^2 ,$$

(40)

where c is the speed of light in a vacuum. This relationship was first tested by Cockcroft and Walton (1932) using the nuclear transformation

$$^7\text{Li} + p \rightarrow 2\alpha + 17.2\,\text{MeV} ,$$

(41)

and it was found that the decrease in mass in this disintegration process was consistent with the observed release of energy, according to Eq. (40). Recently, relation (40) has been verified (Rainville et al., 2005) to within 0.00004%, using very accurate measurements of the atomic-mass difference, Δm, and the corresponding γ-ray wavelength to determine E, the nuclear binding energy, for isotopes of silicon and sulfur.

It has been emphasized by Wilczek (2005) that approximate QCD calculations (Butler et al., 1993; Aoki et al., 2000; Davies et al., 2004) obtain the observed masses of the neutron, proton and other baryons to an accuracy of within 10%. In these calculations, the assumed constituents, quarks and gluons, are taken to be massless. Wilczek concludes that the calculated masses of the hadrons arise from both the energy stored in the motion of the quarks and the energy of the gluon fields, according to Eq. (40): basically the mass of a hadron arises from internal energy.

Wilzcek (2005) has also discussed the underlying principles giving rise to the internal energy, hence the mass, of a hadron. The nature of the gluon color fields is such that they lead to a runaway growth of the fields surrounding an isolated color charge. In fact all this structure (via virtual gluons) implies that an isolated quark would have an infinite energy associated with it. This is the reason why isolated quarks are not seen. Nature requires these infinities to be essentially cancelled or at least made finite. It does this for hadrons in two ways: either by bringing an antiquark close to a quark (i.e forming a meson) or by bringing three quarks, one of each color, together (i.e. forming a baryon) so that in each case the composite hadron is colorless. However, quantum mechanics prevents the quark and the antiquark of opposite colors or the three quarks of different colors from being placed exactly at the same place. This means that the color fields are not exactly cancelled, although sufficiently it seems to remove the infinities associated with isolated quarks. The distribution of the quark-antiquark pairs or the system of three quarks is described by quantum mechanical wave functions. Many different patterns, corresponding to the various hadrons, occur. Each pattern has a characteristic energy, because the color fields are not entirely cancelled and because the quarks are somewhat localized. This characteristic energy, E, gives the characteristic mass, via Eq. (40), of the hadron.

The above picture, within the framework of the SM, provides an understanding of hadron masses as arising mainly from internal energies associated with the strong color interactions. However, as discussed in Subsection 2.2.3, the masses of the elementary particles of the SM, the leptons, the quarks and the W and Z bosons, are interpreted in a completely different way. A "condensate» called the Higgs scalar field (Englert and Brout, 1964; Higgs, 1964), analogous to the Cooper pairs in a superconducting material, is assumed to exist. This field couples, with an appropriate strength, to each lepton, quark and vector boson and endows an originally massless particle with its physical mass. Thus, the assumption of a Higgs field within the framework of the SM not only adds an extra field but also leads to the introduction of 14 new parameters. Moreover, as pointed out by Lyre (2008), the introduction of the Higgs field in the SM to spontaneously break the U(1) × SU(2)L local gauge symmetry of the electroweak interaction to generate the masses of the W and Z bosons, simply corresponds mathematically to putting in "by hand» the masses of the elementary particles of the SM: the so-called Higgs mechanism does not provide any physical explanation for the origin of the masses of the leptons, quarks and the W and Z bosons.

In the CGM (Robson, 2005; 2011a), the elementary particles of the SM have a substructure, consisting of massless rishons and/or antirishons bound together by strong color interactions, mediated by massless neutral hypergluons. This model is very similar to that of the SM in which the quarks and/or antiquarks are bound together by strong color interactions, mediated by massless neutral gluons, to form hadrons. Since, as discussed above, the mass of a hadron arises mainly from the energy of its constituents, the CGM suggests (Robson, 2009) that the mass of a lepton, quark or vector boson arises entirely from the energy stored in the motion of its constituent rishons and/or antirishons and the energy of the color hypergluon fields, E, according to Eq. (40). A corollary of this idea is: if a particle has mass, then it is composite. Thus, unlike the SM, the GM provides a unified description of the origin of all mass.

MASS HIERARCHY OF LEPTONS AND QUARKS

Table 8 shows the observed masses of the charged leptons together with the estimated masses of the quarks: the masses of the neutral leptons have not yet been determined but are known to be very small. Although the mass of a single quark is a somewhat abstract idea, since quarks do not exist as particles independent of the environment around them, the masses of the quarks may be inferred from mass differences between hadrons of similar composition. The strong binding within hadrons complicates the issue to some extent but rough estimates of the quark masses have been made (Veltman, 2003), which are sufficient for our purposes.

Table 8: Masses of leptons and quarks

Charge	0	-1	$+\frac{2}{3}$	$-\frac{1}{3}$
Generation 1	ν_e	e^-	u	d
Mass	< 0.3 eV	0.511 MeV	5 MeV	10 MeV
Generation 2	ν_μ	μ^-	c	s
Mass	< 0.3 eV	106 MeV	1.3 GeV	200 MeV
Generation 3	ν_τ	τ^-	t	b
Mass	< 0.3 eV	1.78 GeV	175 GeV	4.5 GeV

The SM is unable to provide any understanding of either the existence of the three generations of leptons and quarks or their mass hierarchy indicated in Table 8; whereas the CGM suggests that both the existence and mass hierarchy of these three generations arise from the substructures of the leptons and quarks (Robson, 2009; 2011a).

Subsection 3.3 describes the proposed rishon and/or antirishon substructures of the three generations of leptons and quarks and indicates how the pattern of the first generation is followed by the second and third generations. Section 4.1 discusses the origin of mass in composite particles and postulates that the mass of a lepton or quark arises from the energy of its constituents.

In the CGM it is envisaged that the rishons and/or antirishons of each lepton or quark are very strongly localized, since to date there is no direct evidence for any substructure of these particles. Thus the constituents are expected to be distributed according to quantum mechanical wave functions, for which the product wave function is significant for only an extremely small volume of space so that the corresponding color fields are almost cancelled. The constituents of each lepton or quark are localized within a very small volume of space by strong color interactions acting between the colored rishons and/or antirishons. We call these intra-fermion color interactions. However, between any two leptons and/or quarks there will be a residual interaction, arising from the color interactions acting between the constituents of one fermion and the constituents of the other fermion. We refer to these interactions as inter-fermion color interactions. These will be associated with the gravitational interaction and are discussed in the next subsection.

The mass of each lepton or quark corresponds to a characteristic energy primarily associated with the intra-fermion color interactions. It is expected that the mass of a composite particle will be greater if the degree of localization of its constituents is smaller (i.e. the constituents are on average more widely separated). This is a consequence of the nature of the strong color interactions, which are assumed to possess the property of "asymptotic freedom" (Gross and Wilczek, 1973; Politzer, 1973), whereby the color interactions become stronger for larger separations of the color charges. In addition, it should be noted that the electromagnetic interactions between charged T-rishons or between charged $\overline{T}$-antirishons will also cause the degree of localization of the constituents to be smaller causing an increase in mass. There is some evidence for the above expectations. The electron consists of three $\overline{T}$-antirishons, while the electron neutrino consists of three neutral $\overline{V}$-antirishons. Neglecting the electric charge carried by the $\overline{T}$-antirishon, it is expected that the electron and its neutrino would have identical masses, arising from the similar intra-fermion color interactions. However, it is anticipated that the electromagnetic interaction in the electron case will cause the $\overline{T}$-antirishons to be less localized than the $\overline{V}$-antirishons constituting the electron neutrino so that the electron will have a substantially greater characteristic energy and hence a greater mass than the electron neutrino, as observed. This large difference in the masses of the e − and ve leptons (see Table 8) indicates that the mass of a particle is extremely sensitive to the degree of localization of its constituents. Similarly, the up, charmed and top quarks, each containing two charged T-rishons, are expected to have a greater mass than their weak isospin partners, the down, strange and bottom quark, respectively, which contain only a single charged $\overline{T}$-antirishon. This is true provided one takes into account quark mixing (Evans and Robson, 2006) in the case of the up and down quarks, although Table 8 indicates that the down quark is more massive than the up quark, leading to the neutron having a greater mass than the proton. This is understood within the framework of the GM since due to the manner in which quark masses are estimated, it is the weak eigenstate quarks, whose masses are given in Table 8. Since each succeeding generation is significantly more massive than the previous one, any mixing will noticeably increase the mass of a lower generation quark. Thus the weak eigenstate d'-quark, which contains about 5% of the mass eigenstate s-quark, is expected to be significantly more massive than the mass eigenstate d-quark (see Subsection 3.2). We shall now discuss the mass hierarchy of the three generations of leptons and quarks in more detail.

It is envisaged that each lepton of the first generation exists in an antisymmetric three-particle color state, which physically assumes a quantum mechanical triangular distribution of the three differently colored identical rishons (or antirishons) since each of the three color interactions between pairs of rishons (or antirishons) is expected to be strongly attractive (Halzen and Martin, 1984). As indicated above, the charged leptons are predicted to have larger masses than the neutral leptons, since the electromagnetic interaction in the charged leptons will cause their constituent rishons (or antirishons) to be less localized than those constituting the uncharged leptons, leading to a substantially greater characteristic energy and a correspondingly greater mass.

In the CGM, each quark of the first generation is a composite of a colored rishon and a colorless rishon-antirishon pair, ($T\overline{V}$) or a ($V\overline{T}$) (see Table 5). This color charge structure of the quarks is expected to lead to a quantum mechanical linear distribution of the constituent rishons and antirishons, corresponding to a considerably larger mass than that of the leptons, since the constituents of the quarks are less localized. This is a consequence of the character (i.e. attractive or repulsive) of the color interactions at small distances (Halzen and Martin, 1984). The general rules for small distances of separation are:

1.   rishons (or antirishons) of like colors (or anticolors) repel: those having different colors (or anticolors) attract, unless their colors (or anticolors) are interchanged and the two rishons (or antirishons) do not exist in an antisymmetric color state (e.g. as in the case of leptons);

2.   rishons and antirishons of opposite colors attract but otherwise repel.

Furthermore, the electromagnetic interaction occurring within the up quark, leads one to expect it to have a larger mass than that of the down quark.

Each lepton of the second generation is envisaged to basically exist in an antisymmetric three-particle color state, which physically assumes a quantum mechanical triangular distribution of the three differently colored identical rishons (or antirishons), as for the corresponding lepton of the first generation. The additional colorless rishon-antirishon pair, (VU^-) or (UV^-), is expected to be attached externally to this triangular distribution, leading quantum mechanically to a less localized distribution of the constituent rishons and/or antirishons, so that the lepton has a significantly larger mass than its corresponding first generation lepton.

Each quark of the second generation has a similar structure to that of the corresponding quark of the first generation, with the additional colorless rishon-antirishon pair, (VU^-) or (UV^-), attached quantum mechanically so that the whole rishon structure is essentially a linear distribution of the constituent rishons and antirishons. This structure is expected to be less localized, leading to a larger mass relative to that of the corresponding quark of the first generation, with the charmed quark having a greater mass than the strange quark, arising from the electromagnetic repulsion of its constituent two charged T-rishons.

Each lepton of the third generation is considered to basically exist in an antisymmetric three-particle color state, which physically assumes a quantum mechanical triangular distribution of the three differently colored identical rishons (or antirishons), as for the corresponding leptons of the first and second generations. The two additional colorless rishon-antirishon pairs, $(V\bar{U})(V\bar{U})$, $(V\bar{U})(U\bar{V})$ or $(U\bar{V})(U\bar{V})$, , are expected to be attached externally to this triangular distribution, leading to a considerably less localized quantum mechanical distribution of the constituent rishons and/or antirishons, so that the lepton has a significantly larger mass than its corresponding second generation lepton.

Each quark of the third generation has a similar structure to that of the first generation, with the additional two rishon-antirishon pairs (VU^-) and/or (UV^-) attached quantum mechanically so that the whole rishon structure is essentially a linear distribution of the constituent rishons and antirishons. This structure is expected to be even less localized, leading to a larger mass relative to that of the corresponding quark of the second generation, with the top quark having a greater mass than the bottom quark, arising from the electromagnetic repulsion of its constituent two charged T-rishons.

The above is a qualitative description of the mass hierarchy of the three generations of leptons and quarks, based on the degree of localization of their constituent rishons and/or antirishons. However, in principle, it should be possible to calculate the actual masses of the leptons and quarks by carrying out QCD-type computations, analogous to those employed for determining the masses of the proton and other baryons within the framework of the SM (Butler et al., 1993; Aoki et al., 2000; Davies et al., 2004).

Origin of Gravity

Robson (2009) proposed that the residual interaction, arising from the incomplete cancellation of the inter-fermion color interactions acting between the rishons and/or antirishons of one colorless particle and those of another colorless particle, may be identified with the usual gravitational interaction, since it has several properties associated with that interaction: universality, infinite range and very weak strength. Based upon this earlier conjecture, Robson (2011a) has presented a quantum theory of gravity, described below, leading approximately to Newton's law of universal gravitation.

The mass of a body of ordinary matter is essentially the total mass of its constituent electrons, protons and neutrons. It should be noted that these masses will depend upon the environment in which the particle exists: e.g. the mass of a proton in an atom of helium will differ slightly from that of a proton in an atom of lead. In the CGM, each of these three particles is considered to be colorless. The electron is composed of three T^--antirishons, each carrying a different anticolor charge, antired, antigreen or antiblue. Both the proton and neutron are envisaged (as in the SM) to be composed of three quarks, each carrying a different color charge, red, green or blue. All three particles are assumed to be essentially in a three-color antisymmetric state, so that their behavior with respect to the strong color interactions is expected basically to be the same. This similar behavior suggests that the proposed residual interaction has several properties associated with the usual gravitational interaction.

Firstly, the residual interaction between any two of the above colorless particles, arising from the inter-fermion color interactions, is predicted to be of a universal character.

Secondly, assuming that the strong color fields are almost completely cancelled at large distances, it seems plausible that the residual interaction, mediated by massless hypergluons, should have an infinite range, and tend to zero as $1/r^2$. These properties may be attributed to the fact that the constituents of each colorless particle are very strongly localized so that the strength of the residual interaction is extremely weak, and consequently the hypergluon self-interactions are also practically negligible. This means that one may consider the color interactions using a perturbation approach: the residual color interaction is the sum of all the two-particle color charge interactions, each of which may be treated perturbatively, i.e. as a single hypergluon exchange. Using the color factors (Halzen and Martin, 1984) appropriate for the SU(3) gauge field, one finds that the residual color interactions between any two colorless particles (electron, neutron or proton) are each attractive.

Since the mass of a body of ordinary matter is essentially the total mass of its constituent electrons, neutrons and protons, the total interaction between two bodies of masses, m_1 and m_2, will be the sum of all the two-particle contributions so that the total interaction will be proportional to the product of these two masses, $m_1 m_2$, provided that each two-particle interaction contribution is also proportional to the product of the masses of the two particles.

This latter requirement may be understood if each electron, neutron or proton is considered physically to be essentially a quantum mechanical triangular distribution of three differently colored rishons or antirishons. In this case, each particle may be viewed as a distribution of three color charges throughout a small volume of space with each color charge having a certain probability of being at a particular point, determined by its corresponding color wave function. The total residual interaction between two colorless particles will then be the sum of all the intrinsic interactions acting between a particular triangular distribution of one particle with that of the other particle.

Now the mass m of each colorless particle is considered to be given by $m = E/c^2$, where E is a characteristic energy, determined by the degree of localization of its constituent rishons and/or antirishons. Thus the significant volume of space occupied by the triangular distribution of the three differently colored rishons or antirishons is larger the greater the mass of the particle. Moreover, due to antiscreening effects (Gross and Wilczek, 1973; Politzer, 1973) of the strong color fields, the average strength of the color charge within each unit volume of the larger localized volume of space will be increased. If one assumes that the mass of a particle is proportional to the integrated sum of the intra-fermion interactions within the significant volume of space occupied by the triangular distribution, then the total residual interaction between two such colorless particles will be proportional to the product of their masses.

Thus the residual color interaction between two colorless bodies of masses, m1 and m2, is proportional to the product of these masses and moreover is expected to depend approximately as the inverse square of their distance of separation r, i.e. as $1/r^2$, in accordance with Newton's law of universal gravitation. The approximate dependence on the inverse square law is expected to arise from the effect of hypergluon self-interactions, especially for large separations. Such deviations from an inverse square law do not occur for electromagnetic interactions, since there are no corresponding photon self-interactions.

Mixed-Quark States in Hadrons

As discussed in Subsection 3.2 the GM postulates that hadrons are composed of weak eigenstate quarks rather than mass eigenstate quarks as in the SM. This gives rise to several important consequences (Evans and Robson, 2006; Morrison and Robson, 2009; Robson, 2011b; 2011c).

Firstly, hadrons composed of mixed-quark states might seem to suggest that the electromagnetic and strong interaction processes between mass eigenstate hadron components are not consistent with the fact that weak interaction processes occur between weak eigenstate quarks. However, since the electromagnetic and strong interactions are flavor independent: the down, strange and bottom quarks carry the same electric and color charges so that the weak eigenstate quarks have the same magnitude of electric and color charge as the mass eigenstate quarks. Consequently, the weak interaction is the only interaction in which the quark-mixing phenomenon can be detected.

Secondly, the occurrence of mixed-quark states in hadrons implies the existence of higher generation quarks in hadrons. In particular, the GM predicts that the proton contains $\approx 1.7\%$ of strange quarks, while the neutron having two d$'$-quarks contains $\approx 3.4\%$ of strange quarks. Recent experiments (Maas et al., 2005; Armstrong et al, 2005) have provided some evidence for the existence of strange quarks in the proton. However, to date the experimental data are compatible with the predictions of both the GM and the SM ($\ll 1.7\%$).

Thirdly, the presence of strange quarks in nucleons explains why the mass of the neutron is greater than the mass of a proton, so that the proton is stable. This arises because the mass of the weak eigenstate d$'$-quark is larger than the mass of the u-quark, although the mass eigenstate d-quark is expected to be smaller than that of the u-quark, as discussed in the previous section. Another consequence of the presence of mixed-quark states in hadrons is that mixed-quark states may have mixed parity. In the CGM the constituents of quarks are rishons and/or antirishons.

If one assumes the simple convention that all rishons have positive parity and all their antiparticles have negative parity, one finds that the down and strange quarks have opposite intrinsic parities, according to the proposed structures of these quarks in the CGM: the d-quark (see Table 5) consists of two rishons and one antirishon ($P_d = -1$), while the s-quark (see Table 6) consists of three rishons and two antirishons (Ps = +1). The u-quark consists of two rishons and one antirishon so that $P_u = -1$, and the antiparicles of these three quarks have the corresponding opposite parities: $P_{-d} = +1$, $P_{-s} = -1$ and $P_{-u} = +1$.

In the SM the intrinsic parity of the charged pions is assumed to be $P\pi = -1$. This result was established by Chinowsky and Steinberger (1954), using the capture of negatively charged pions in deuterium to form two neutrons, and led to the overthrow of the conservation of both parity (P) and charge-conjugation (C) (Lee and Yang, 1956; Wu et al., 1957; Garwin et al., 1957; Friedman and Telegdi, 1957) and later combined CP conservation (Christenson et al., 1964). Recently, Robson (2011b) has demonstrated that this experiment is also compatible with the mixed-parity nature of the $\pi -$ predicted by the CGM: $\approx (0.95P_d + 0.05Ps)$, with $P_d = -1$ and $P_s = +1$. This implies that the original determination of the parity of the negatively charged pion is not conclusive, if the pion has a complex substructure as in the CGM. Similarly, Robson (2011c) has shown that the recent determination (Abouzaid et al., 2008) of the parity of the neutral pion, using the double Dalitz decay $\pi^0 \rightarrow e^+ e^- e^+ e^-$ is also compatible with the mixed-parity nature of the neutral pion predicted by the CGM.

This new concept of mixed-parity states in hadrons, based upon the existence of weak eigenstate quarks in hadrons and the composite nature of the mass eigenstate quarks, leads to an understanding of CP symmetry in nature. This is discussed in the following subsection.

CP Violation in the $K^0 - \bar{K}^0$ System

Gell-Mann and Pais (1955) considered the behavior of neutral particles under the charge-conjugation operator C. In particular they considered the K^0 meson and realized that unlike the photon and the neutral pion, which transform into themselves under the C operator so that they are their own antiparticles, the antiparticle of the K^0 meson (strangeness $S = +1$), K^0, was a distinct particle, since it had a different strangeness quantum number ($S = -1$). They concluded that the two neutral mesons, K^0 and $\bar{K}^0$, are degenerate particles that exhibit unusual properties, since they can transform into each other via weak interactions such as

$$K^0 \rightleftharpoons \pi^+ \pi^- \rightleftharpoons \bar{K}^0.$$

(42)

In order to treat this novel situation, Gell-Mann and Pais suggested that it was more convenient to employ different particle states, rather than K^0 and $\bar{K}^0$, to describe neutral kaon decay. They suggested the following representative states:

$$K_1^0 = (K^0 + \bar{K}^0)/\sqrt{2}, \quad K_2^0 = (K^0 - \bar{K}^0)/\sqrt{2},$$

(43)

and concluded that these particle states must have different decay modes and lifetimes. In particular they concluded that K_1^0 could decay to two charged pions, while K_2^0 would have a longer lifetime and more complex decay modes. This conclusion was based upon the conservation of C in the weak interaction processes: both K_1^0 and the $\pi^+\pi^-$ system are even (i.e. C = +1) under the C operation.

The particle-mixing theory of Gell-Mann and Pais was confirmed in 1957 by experiment, in spite of the incorrect assumption of C invariance in weak interaction processes. Following the discovery in 1957 of both C and P violation in weak interaction processes, the particle-mixing theory led to a suggestion by Landau (1957) that the weak interactions may be invariant under the combined operation CP.

Landau's suggestion implied that the Gell-Mann–Pais model of neutral kaons would still apply if the states, K_1^0 and K_2^0, were eigenstates of CP with eigenvalues +1 and −1, respectively. Since the charged pions were considered to have intrinsic parity $P_\pi = -1$, it was clear that only the K_1^0 state could decay to two charged pions, if CP was conserved.

The suggestion of Landau was accepted for several years since it nicely restored some degree of symmetry in weak interaction processes. However, the surprising discovery (Christenson et al., 1964) of the decay of the long-lived neutral K 0 meson to two charged pions led to the conclusion that CP is violated in the weak interaction. The observed violation of CP conservation turned out to be very small ($\approx$ 0.2%) compared with the maximal violations ($\approx$ 100%) of both P and C conservation separately. Indeed the very smallness of the apparent CP violation led to a variety of suggestions explaining it in a CP-conserving way (Kabir, 1968; Franklin, 1986). However, these efforts were unsuccessful and CP violation in weak interactions was accepted.

An immediate consequence of this was that the role of K_1^0 (CP = +1) and K_2^0 (CP = −1), defined in Eqs. (43), was replaced by two new particle states, corresponding to the short-lived (K_S^0) and long-lived (K_L^0) neutral kaons:

$$K_S^0 = (K_1^0 + \epsilon K_2^0)/(1 + |\epsilon|^2)^{\frac{1}{2}}, \quad K_L^0 = (K_2^0 + \epsilon K_1^0)/(1 + |\epsilon|^2)^{\frac{1}{2}},$$
(44)

where the small complex parameter is a measure of the CP impurity in the eigenstates K_S^0 and K_L^0. This method of describing CP violation in the Standard Model (SM), by introducing mixing of CP eigenstates, is called 'indirect CP violation'. It is essentially a phenomenological approach with the parameter to be determined by experiment.

Another method of introducing CP violation into the SM was proposed by Kobayashi and Maskawa (1973). By extending the idea of 'Cabibbo mixing' (see Subsection 2.2.3) to three generations, they demonstrated that this allowed a complex phase to be introduced into the quark-mixing (CKM) matrix, permitting CP violation to be directly incorporated into the weak interaction. This phenomenological method has within the framework of the SM successfully accounted for both the indirect CP violation discovered by Christenson et al. in 1964 and the "direct CP violation" related to the decay processes of the neutral kaons (Kleinknecht, 2003). However, to date, the phenomenological approach has not been able to provide an a priori reason for CP violation to occur nor to indicate the magnitude of any such violation.

Recently, Morrison and Robson (2009) have demonstrated that the indirect CP violation observed by Christenson et al. (1964) can be described in terms of mixed-quark states in hadrons. In addition, the rate of the decay of the K^0_L meson relative to the decay into all charged modes is estimated accurately in terms of the Cabibbo-mixing angle.

In the CGM the K^0 and $\bar{K}^0$ mesons have the weak eigenstate quark structures $[d' \bar{s}']$ and $[s' \bar{d}']$, respectively. Neglecting the very small mixing components arising from the third generation, Morrison and Robson show that the long-lived neutral kaon, K^0_L, exists in a CP = -1 eigenstate as in the SM. On the other hand, the charged 2π system:

$$\pi^+\pi^- = [u\bar{d}'][d'\bar{u}]$$
$$= [u\bar{d}][d\bar{u}]\cos^2\theta_c + [u\bar{s}][s\bar{u}]\sin^2\theta_c + [u\bar{s}][d\bar{u}]\sin\theta_c\cos\theta_c$$
$$+[u\bar{d}][s\bar{u}])\sin\theta_c\cos\theta_c .$$

$$(45)$$

For the assumed parities (see Subsection 4.4) of the quarks and antiquarks involved in Eq. (45), it is seen that the first two components are eigenstates of CP = +1, while the remaining two components [u$\bar{s}$][d$\bar{u}$] and [u$\bar{d}$][s$\bar{u}$], with amplitude $\sin\theta_c \cos\theta_c$ are not individually eigenstates of CP. However, taken together, the state ([u$\bar{s}$][d$\bar{u}$]+[u$\bar{d}$][s$\bar{u}$]) is an eigenstate of CP with eigenvalue CP = -1. Taking the square of the product of the amplitudes of the two components comprising the CP = -1 eigenstate to be the "joint probability» of those two states existing together simultaneously, one can calculate that this probability is given by $(\sin\theta_c \cos\theta_c)4 = 2.34 \times 10^{-3}$, using $\cos\theta_c = 0.9742$ (Amsler et al., 2008). Thus, the existence of a small component of the $\pi^+\pi^-$ system with eigenvalue CP = -1 indicates that the K^0_L meson can decay to the charged 2π system without violating CP conservation. Moreover, the estimated decay rate is in good agreement with experimental data (Amsler et al., 2008).

SUMMARY AND FUTURE PROSPECTS

The GM, which contains fewer elementary particles (27 counting both particles and antiparticles and their three different color forms) and only two fundamental interactions (the electromagnetic and strong color interactions), has been presented as a viable simpler alternative to the SM (61 elementary particles and four fundamental interactions).

In addition, the GM has provided new paradigms for particle physics, which have led to a new understanding of several phenomena not addressed by the SM. In particular, (i) the mass of a particle is attributed to the energy content of its constituents so that there is no requirement for the Higgs mechanism; (ii) the mass hierarchy of the three generations of leptons and quarks is described by the degree of localization of their constituent rishons and/or antirishons; (iii) gravity is interpreted as a quantum mechanical residual interaction of the strong color interaction, which binds rishons and/or antirishons together to form all kinds of matter and (iv) the decay of the long-lived neutral kaon is understood in terms of mixed-quark states in hadrons and not CP violation.

The GM also predicts that the mass of a free neutron is greater than the mass of a free proton so that the free proton is stable. In addition, the model predicts the existence of higher generation quarks in hadrons, which in turn predicts mixed-parity states in hadrons. Further experimentation is required to verify these predictions and thereby strengthen the Generation Model.

REFERENCES

1. Abouzaid, E. et al. (2008), Determination of the Parity of the Neutral Pion via its Four-Electron Decay, Physical Review Letters, Vol. 100, No. 18, 182001 (5 pages).

2. Aitchison I.J.R. and Hey, A.J.G. (1982), Gauge Theories in Particle Physics (Adam Hilger Ltd, Bristol).

3. Amsler, C. et al. (2008), Summary Tables of Particle Properties, Physics Letters B, Vol. 667, Nos. 1-5, pp. 31-100.

4. Aoki, S. et al. (2000), Quenched Light Hadron Spectrum, Physical Review Letters, Vol. 84, No. 2, pp. 238-241.

5. Armstrong, D.S. et al. (2005), Strange-Quark Contributions to Parity-Violating Asymmetries in the Forward G0 Electron-Proton Scattering Experiment, Physical Review Letters, Vol. 95, No. 9, 092001 (5 pages).

6. Arnison, G. et al. (1983), Experimental Observation of Isolated Large Transverse Energy Electrons with Associated Missing Energy, Physics Letters B, Vol 122, No. 1, pp. 103-116.

7.   Banner, M. et al. (1983), Observation of Single Isolated Electrons of High Transverse Momentum in Events with Missing Transverse Energy at the CERN pp Collider, Physics Letters B, Vol. 122, Nos. 5-6, pp. 476-485.

8.   Bloom, E.D. et al. (1969), High-Energy Inelastic e − p Scattering at 60 and 100 , Physical Review Letters, Vol. 23, No. 16, pp. 930-934.

9.   Breidenbach, M. et al. (1969), Observed Behavior of Highly Inelastic Electron-Proton Scattering, Physical Review Letters, Vol. 23, No. 16, pp. 935-939.

10. Butler, F. et al. (1993), Hadron Mass Predictions of the Valence Approximation to Lattice QCD, Physical Review Letters, Vol. 70, No. 19, pp. 2849-2852.

11. Cabibbo, N. (1963), Unitary Symmetry and Leptonic Decays, Physical Review Letters, Vol. 10, No. 12, pp. 531-533.

12. Casalbuoni, R. and Gatto, R. (1980), Subcomponent Models of Quarks and Leptons, Physics Letters B, Vol. 93, Nos. 1-2, pp. 47-52.

13. Chinowsky, W. and Steinberger, J. (1954), Absorption of Negative Pions in Deuterium: Parity of the Pion, Physical Review, Vol. 95, No. 6, pp. 1561-1564.

14. Christenson, J.H. et al. (1964), Evidence for the 2π Decay of the K^0_2 Meson, Physical Review Letters, Vol. 13, No. 4, pp. 138-140.

15. Cockcroft, J. and Walton, E. (1932), Experiments with High Velocity Positive Ions. II. The Disintegration of Elements by High Velocity Protons, Proceedings of the Royal Society of London, Series A, Vol. 137, No. 831, pp. 239-242.

16. Davies, C.T.H. et al. (2004), High-Precision Lattice QCD Confronts Experiment, Physical Review Letters, Vol. 92, No. 2, 022001 (5 pages).

17. Einstein, A. (1905), Ist die Trägheit eines Körpers von seinem Energieinhalt abhängig, Annalen der Physik, Vol. 18, No. 13, pp. 639-641.

18. Englert, F. and Brout, R. (1964), Broken Symmetry and the Mass of Gauge Vector Bosons, Physical Review Letters, Vol. 13, No. 9, pp. 321-323.

19. Evans, P.W. and Robson, B.A. (2006), Comparison of Quark Mixing in the Standard and Generation Models, International Journal of Modern Physics E, Vol. 15, No 3, pp. 617-625.

20. Franklin, A. (1986), The Neglect of Experiment (Cambridge University Press, Cambridge, U.K.).

21. Friedman, J.I. and Telegdi, V.L. (1957), Nuclear Emulsion Evidence for Parity Nonconservation in the Decay Chain $\pi^+ - \mu^+ - e^+$, Physical Review, Vol. 105, No. 5, pp. 1681-1682.

22. Garwin, R.L., Lederman, L.M. and Weinrich, M. (1957), Observations of the Failure of Conservation of Parity and Charge Conjugation in Meson Decays: the Magnetic Moment of the Free Muon, Physical Review, Vol. 105, No. 4, pp. 1415-1417.

23. Gell-Mann, M. and Ne'eman, Y. (1964), The Eightfold Way, (Benjamin, New York).

24. Gell-Mann, M. and Pais, A. (1955), Behavior of Neutral Particles under Charge Conjugation, Physical Review, Vol. 97, No. 5, pp. 1387-1389.

25. Glashow, S.L. (1961), Partial-Symmetries of Weak Interactions, Nuclear Physics, Vol. 22, pp. 579-588.

26. Gottfried, K. and Weisskopf, V.F. (1984), Concepts of Particle Physics Vol. 1 (Oxford University Press, New York).

27. Gross, D.J. and Wilczek, F. (1973), Ultraviolet Behavior of Non-Abelian Gauge Theories, Physical Review Letters, Vol. 30, No. 26, pp. 1343-1346.

28. Halzen, F. and Martin, A.D. (1984), Quarks and Leptons: An Introductory Course in Modern Particle Physics (John Wiley and Sons, New York).

29. Harari, H. (1979), A Schematic Model of Quarks and Leptons, Physics Letters B, Vol. 86, No. 1, pp. 83-86.

30. Harari, H. and Seiberg, N. (1981), A Dynamical Theory for the Rishon Model, Physics Letters B, Vol.98, No. 4, pp. 269-273.

31. Hasert et al. (1973), Observation of Neutrino-Like Interactions without Muon or Electron in the Gargamelle Neutrino Experiment, Physics Letters B, Vol. 46, No. 1, pp. 138-140.

32. Hasert et al. (1974), Observation of Neutrino-Like Interactions without Muon or Electron in the Gargamelle Neutrino Experiment, Nuclear Physics B, Vol. 73, No. 1, pp. 1-22.

33. Higgs, P.W. (1964), Broken Symmetries and the Masses of Gauge Bosons, Physical Review Letters, Vol. 13, No. 16, pp. 508-509.

34. Kabir, P.K. (1968), The CP Puzzle: Strange Decays of the Neutral Kaon (Academic Press, London).

35. Kleinknecht, K. (2003), Uncovering CP Violation: Experimental Clarification in the Neutral K Meson and B Meson Systems (Springer, Berlin).

36. Kobayashi, M. and Maskawa, T. (1973), CP-Violation in Renormalizable Theory of Weak Interaction, Progress of Theoretical Physics, Vol. 49, No. 2, pp. 652-657.

37. Landau, L.D. (1957), On the Conservation Laws for Weak Interactions, Nuclear Physics, Vol. 3, No.1, pp. 127-131.

38. Lee, T.D. and Yang, C.N. (1956), Question of Parity Conservation in Weak Interactions, Physical Review, Vol. 104, No. 1, pp. 254-258.

39. Lyre, H. (2008), Does the Higgs Mechanism Exist?, International Studies in the Philosophy of Science, Vol. 22, No. 2, pp. 119-133.

40. Mass, F.E. et al., Evidence for Strange-Quark Contributions to the Nucleon's Form Factors at Q2 = 0.108 (GeV/c) 2, Physical Review Letters, Vol. 94, No. 15, 152001 (4 pages).

41. Morrison, A.D. and Robson, B.A. (2009), 2π Decay of the K0L Meson without CP Violation, International Journal of Modern Physics E, Vol. 18, No. 9, pp. 1825-1830.

42. Politzer, H.D. (1973), Reliable Perturbative Results for Strong Interactions, Physical Review Letters, Vol. 30, No. 26, pp. 1346-1349.

43. Rainville, S. et al. (2005), World Year of Physics: A Direct Test of E = mc^2 , Nature, Vol. 438, pp. 1096-1097.

44. Robson, B.A. (2002), A Generation Model of the Fundamental Particles, International Journal of Modern Physics E, Vol. 11, No. 6, pp. 555-566.

45. Robson, B.A. (2004), Relation between Strong and Weak Isospin, International Journal of Modern Physics E, Vol. 13, No. 5, pp. 999-1018.

46. Robson, B.A. (2005), A Generation Model of Composite Leptons and Quarks, International Journal of Modern Physics E, Vol. 14, No. 8, pp. 1151-1169.

47. Robson, B.A. (2008), The Generation Model and the Electroweak Connection, International Journal of Modern Physics E, Vol. 17, No. 6, pp. 1015-1030.

48. Robson, B.A. (2009), The Generation Model and the Origin of Mass, International Journal of Modern Physics E, Vol. 18, No. 8, pp. 1773-1780.

49. Robson, B.A. (2011a), A Quantum Theory of Gravity based on a Composite Model of Leptons and Quarks, International Journal of Modern Physics E, Vol. 20, No. 3, pp. 733-745.

50. Robson, B.A. (2011b), Parity of Charged Pions, International Journal of Modern Physics E, Vol. 20, No. 8, pp. 1677-1686.

51. Robson, B.A. (2011c), Parity of Neutral Pion, International Journal of Modern Physics E, Vol. 20, No. 9, pp. 1961-1965.

52. Salam, A. (1968) in Elementary Particle Physics (Proceedings of the 8th Nobel Symposium), ed. Svartholm, N. (Almqvist and Wiksell, Stockholm), p. 367.

53. Shupe, M.A. (1979), A Composite Model of Leptons and Quarks, Physics Letters B, Vol. 86, No. 1, pp. 87-92.

54. Squires, E.J. (1980), QDD-a Model of Quarks and Leptons, Physics Letters B, Vol. 94, No. 1, pp. 54-56.

55. Squires, E.J. (1981), Some Comments on the Three-Fermion Composite Quark and Lepton Model, Journal of Physics G, Vol. 7, No. 4, pp. L47-L49.

56. Hooft, G. (1971a), Renormalization of Massless Yang-Mills Fields, Nuclear Physics B, Vol. 33, No. 1, pp. 173-199.

57. t'Hooft, G. (1971b), Renormalizable Lagrangians for Massive Yang-Mills Fields, Nuclear Physics B, Vol. 35, No. 1, pp. 167-188.

58. Veltman, M. (2003), Facts and Mysteries in Elementary Particle Physics, (World Scientific Publishing Company, Singapore).

59. Weinberg, S. (1967), A Model of Leptons, Physical Review Letters, Vol. 19, No. 21, pp. 1264-1266.

60. Wilczek, F. (2005) In Search of Symmetry Lost, Nature, Vol. 433, No. 3, pp. 239-247.

61. Wu, C.S. et al. (1957), Experimental Test of Parity Conservation in Beta Decay, Physical Review, Vol. 105, No. 4, pp. 1413-1415

Chapter 2

STUDY OF "RADIATION EFFECTS OF NUCLEAR HIGH ENERGY PARTICLES" ON ELECTRONIC CIRCUITS AND METHODS TO REDUCE ITS DESTRUCTIVE EFFECTS

Omid Zeynali, Daryoush Masti, Maryam Nezafat, Alireza Mallahzadeh

Islamic Azad University, Dashtestan Branch, Borazjan, Iran

ABSTRACT

This research concerns on (TID), (DD) and (SEE) effects also high energy particles' effects on electronic properties of silicon. It investigates the silicon electronic properties exposed to these particles using a laboratory neutron radiation sources. Some Pieces of a silicon wafer were under neutron radiation at different times and the electrical properties of each one was illustrated by plate resistance measurement and also the strength of the current voltage was simulated by Fluka and MCNP software. Based on these results, authorized limit of silicon tolerance was obtained against high energy neutrons radiation. We put them in the electric furnace under thermal recovery to overcome the unusual behavior of irradiated samples.

INTRODUCTION

By expanding the field of human space and nuclear energy exploration technologies, there are a huge demands on semiconductor components in the nuclear radiation environments. The pieces are used in different parts of spacecraft and nuclear plants. High energy particles such as electrons, protons and neutrons are some other kinds of radiation in nuclear environments [1]. Ionizing radiations emitted from radiation sources are the main part of space radiation environment and nuclear sites, the prevalence of abnormalities on electronic systems and communications satellites and nuclear power plants [2]. For example, the nuclear reactor parameters and pressure, temperature and neutron flux transducers are measured by sensors which the sensors and transducers are used in the electronics circuits and these devices are energetic particles and radiation in the environment. The nuclear radiation has some effects on various electronic circuits and telecommunications, particularly the effects of total ionizing dose (TID) and single event effects (SEE) [3].

THEORY AND BACKGROUND

Ionizing radiations are the radiations which in collision with atoms, have enough energy to separate the electrons orbiting atoms and charged or ionized atoms. High energy ionized particles such as protons, heavy ions and electrons influence on electronic systems. These effects can decrease the efficiency and performance of the system or off the device. So the equipments may face a short lifespan or defect and a huge disadvantage [1]. In reducing the amount of obtained energy of the electronic components a shield can be used [4]. But this method cannot counter single event effects and total ionization forced by high energy ionized particles, so it may cause some defects and probable errors in each part [5]. Semiconductors are the main structure of the electronic components. Semiconductor devices properties after exposure to radiation are so important they the behavior and lifetime of these parts can be predicted, because radiation creates irreversible damage [6]. In this study, it is necessary to provide high energy particle radiation conditions in the laboratory environment. The laboratory neutron radiation sources were used. Neutrons are preferred because; high energy radiation from other particles can also be simulated by it. Creation of crystal defects in semiconductors is mainly due to the improper position of atoms within the network instead of ordinary position. The fast neutrons can be localized cluster like defects in the material, during manufacturing penetrating to other areas and therefore leading to the breakdown of voltage and changing the device leakage current. Irradiating the crystal with high energy neutrons, results a wide range of defects. Such shortcomings have elliptical or nearly spherical shape. Frnkly couples are the simplest defects resulted by fast neutrons. The numbers of Frnkly couples formed per a fast neutron are estimated by KINE CHIN PICE model according to the following relation [5]:

$$v = \langle E_p \rangle / 2E_d$$

(1)

where <ep> is average energy transferred from neutrons collision to hit the atom. <ed> is also the energy needed to move host atoms of the network in its place. Therefore, with the average energy of fast neutron entered the crystal and the radiation time per fast neutrons, the number of Frnkly couples and finally the number of defects can be estimated. Although the actual amount of defects created, greatly less than the value calculated from the formula above, however microscopic amorphous regions will be created in the crystal.

Another reason for the difference in the density of defects with real calculation is the exact amount of energy not determined by neutron experiments.</e</e

LABORATORY ACTIVITIES

The silicon wafer of n (100) type was cut to the sample sizes of 1 cm × 2 cm × 0.7 mm and after the cleaning and cutting; four-point probe resistivity of the samples was measured. To ensure repeatability, every step of the experiment was carried out by a number of, samples and the resistance test of each sample was done in several different spots, and almost the same results were obtained in each group. The measurement result of a sample before irradiation is shown in**Figure 1**. As it can be seen in the figure, the current changes in the amount of impedance (resistance plate) in the range of 450 - 1800 ohms.

The samples were then irradiated. This experiment was done by using a neutron beam with flux control features $\Phi 1 = 2.8 \times 1015$ n·cm−2 and the value $j = 2 \times 108$ n·cm−2·s−1 the samples also had different radiation times at first. In this study, neutron irradiation can be divided into three categories according to their energy as follows: 1) Fast neutrons with energies greater than 10 Kev. 2) Intermediate neutrons (Epithermal) energy of less than 10 Kev and more than 1 ev. 3) Thermal neutrons with energy less than 1 ev.

Fast neutrons have small absorption cross section, it means that the absorption of these neutrons by these elements is little and mainly these neutrons destruct the crystalline semiconductors and cause different defects. Intermediate neutrons make point defects [5]. For better understanding the radiation process, the relative flux of neutrons were estimated in the chamber for three energies above using the MCNP and Fluka software [7,8]. Considering a sample in shield changes neutrons distribution so it was simulated with and without the sample. Estimation result of a sample is shown in **Figure 2**. The samples with very low absorption cross section of semiconductors virtually have "no effect on the chart".

In this study, five samples were selected for irradiation characteristics based on the rate and duration of radiation exposure in accordance with **Table 1**. These values are calculated based on the MCNP software [8]. The lowest dose of E was seen close to the neutron source and has fast neutron flow.

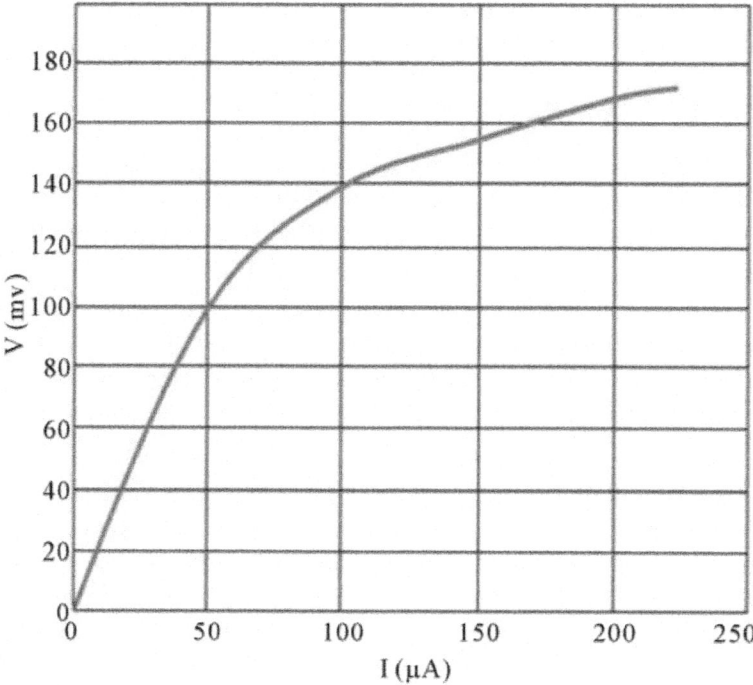

Figure 1: Characteristic of voltage—instance current of group A before the neutron irradiation.

After irradiation, the samples were taken away from the neutron source and after elapsing at least 10 times the half life decay, the activity levels were measured at first. After making sure they are not radioactive, resistivity and voltage-current characteristics of them were measured and plotted again in the new conditions. The four probe device was used to measure resistivity. The results of a four-point probe method for measuring the voltage in terms of current on the sample A, after irradiation are plotted in **Figure 3**. Measurements were repeated several times and in different environments and nearly the same results were obtained. Based on the characteristics, voltage-current relationship was not as previous. Secondly, the average resistance value is about two mega-ohms about 4000 times larger than the amount of measured resistance before radiation.

The increase of ohmic resistance of the sample is the reason of the disassemble in the crystal lattice, and its defects. In order to overcome the unusual behavior of irradiated samples, we put them in the electric furnace

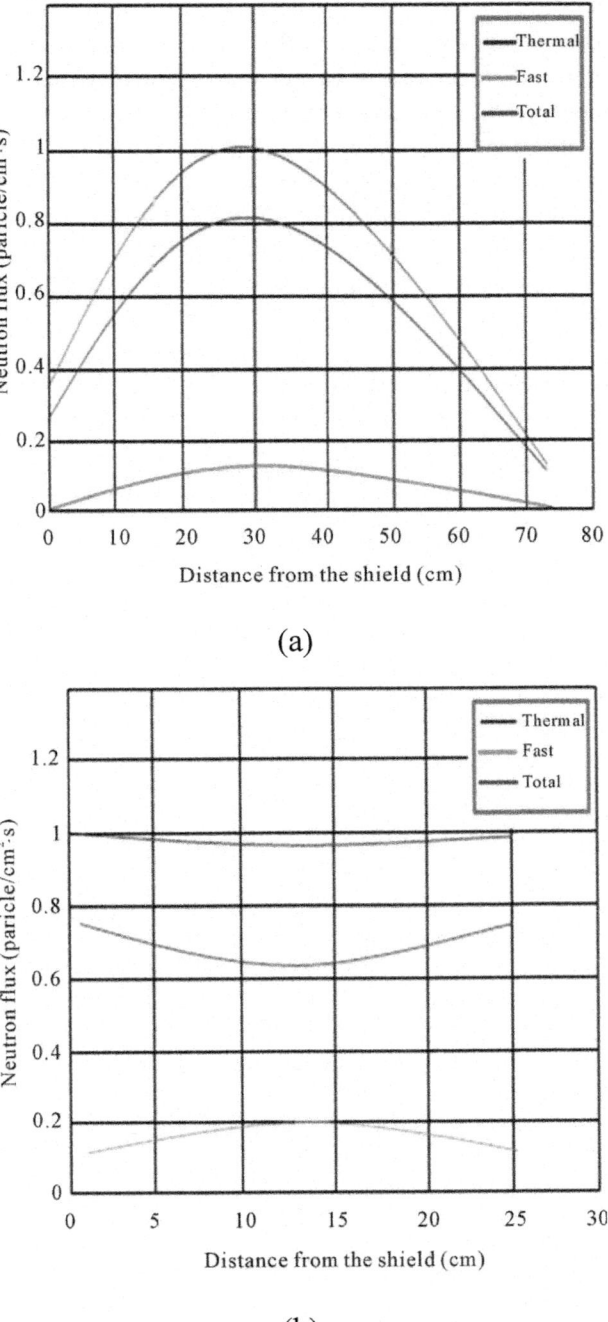

(a)

(b)

Figure 2: Simulation of the neutron flux distribution in the shield using the software in the case of FLUKA in compartment (a) and (b) in no compartment.

with a temperature of 380°C for 10 hours under Thermal (thermal recovery) [9]. Due to the presence of defects in the samples, all of them needed thermal recovery. Intermediate neutrons cause point defects which are resolved by thermal recovery at temperatures of about 400°C. By thermal treatment some voltage-current defects were removed and less radiation in samples returns impedance amount to the amount before radiation. For example, Sample characteristic of thermal treatment is shown in **Figure 4**. In order to remove defects the temperature of 400°C for thermal neutrons, and 750°C for fast neutrons, was necessary to eliminate fast neutrons defects after thermal recovery.

Table 1: characteristics of irradiation samples

Flux rate ($n \cdot cm^{-2} \cdot s^{-1}$)	Irradiation time (hourse)	Sample
10^{13}	1.5	group A
1.8×10^{13}	3.3	group B
4×10^{12}	16.25	group C
10^{13}	42.5	group D
2×10^{11}	4.5	group E

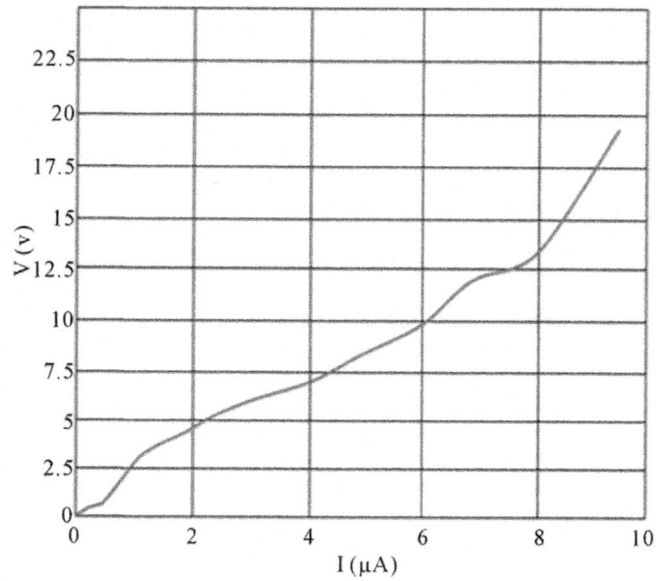

Figure 3: Typical voltage-current characteristics of group A after neutron irradiation.

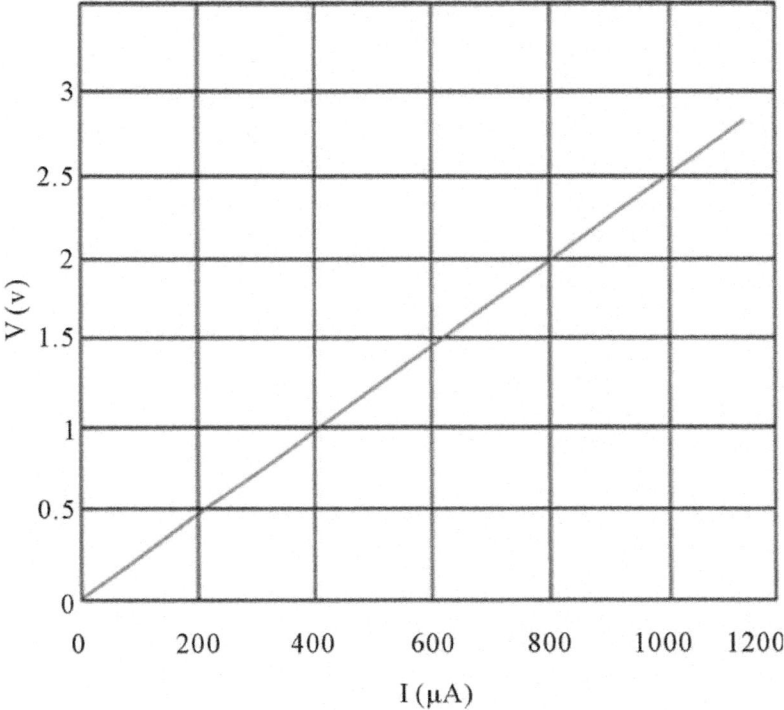

Figure 4: Typical voltage-current characteristics of group A after neutron irradiation and thermal recovery.

The results of measuring the voltage were obtained in terms of current on the four point probe method for samples A, C, D, E after irradiation (results after thermal recovery), the graph of **Figure 5** respectively. As it is observed only sample A is returned back to original state and other samples have significant changes in their resistance.

SIMULATION OF RADIATION HIGH ENERGY PARTIC-LESS' ENVIRONMENT

Energetic ionized particles are classified into three categories [10]: Space belts—Consisting particles such as electrons (to the extent of 30 Mev), protons (to 200Mev), inner belt (450 to 3500 km altitude), and the outer Van Allen belt (8500 to 16,000 km altitude). Spectrum of electrons (500 ev - 20 Mev) in this area for nuclear environment by Fluka software is simulated and calculated [7] and the rate of high energy protons and electrons beam energy, are shown in figures 6 and 7. The high energy protons or electrons produce the network defects in semiconductors crystalline and cause displacement damage [5].

RADIATION EFFECTS ON ELECTRONICS

The nuclear radiation effects are divided into two categories: Cumulative effects and single event effects (SEE). Negative and destructive effects on the electronic components are because of protons and electrons and heavy ions and result to unusual behavior in most electronics [11].

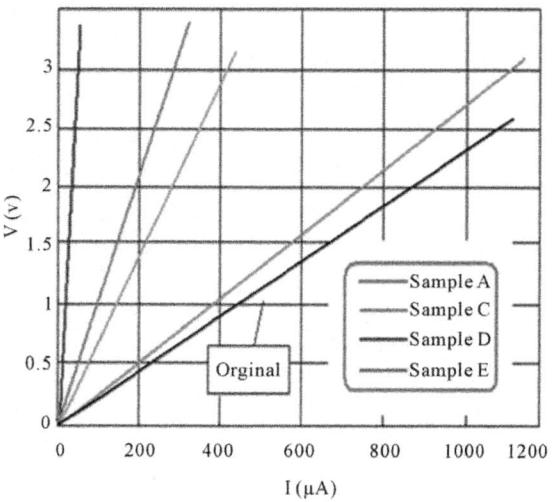

Figure 5: Drawing graphs in a coordinate system in a logarithmic scale to the thermal recovery.

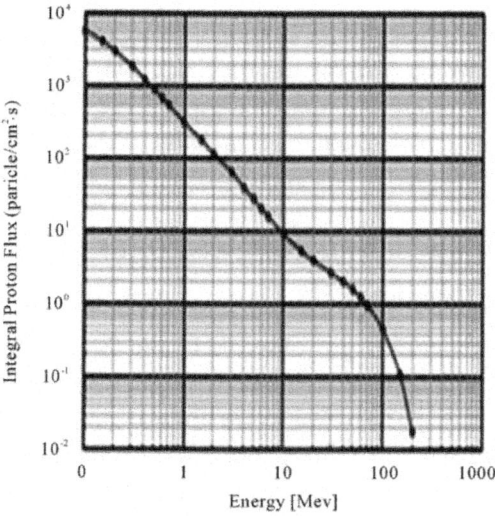

Figure 6: The energy spectrum of trapped protons and electrons, for the calculation of solar proton spectra [1], the Fluka software was used.

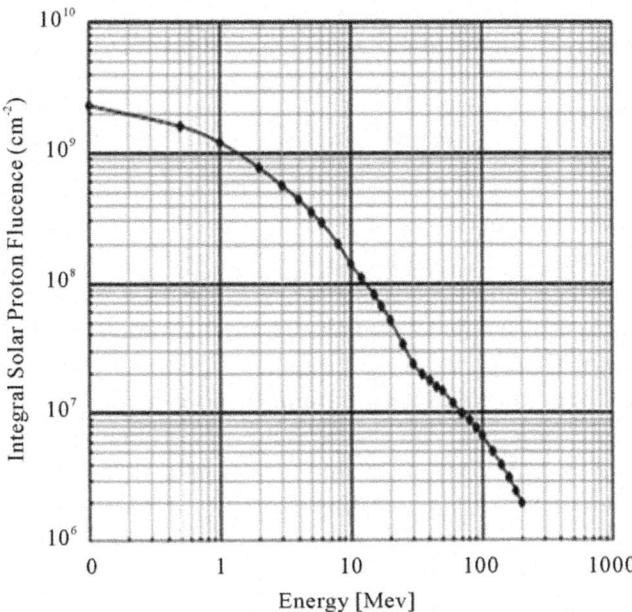

Figure 7: The energy spectrum of the high energy protons.

Cumulative Effects

These effects are derived from the production or activation of microscopic defects in electronic components semiconductors. Cumulative effects are classified in two parts: Total ionization dose (TID) and displacement damage (DD). When a high energy particle collides with mater, there are two major phenomenons [12]: 1) Ionization phenomenon, which is the interaction of particles with the atomic shell electrons. This phenomenon causes hole-electron pairs. Electrons come out due to the greater freedom of movement, and holes trap in the border $SiO2$ as the conductors in electronics assembly are and can eventually cause a change in the electrical specifications. 2) Interaction with the nucleus of an atom by transferring enough energy the atom displacement. Indeed, high energy protons or electrons result in network defects in semiconductors and, leading to the production and displacement damage (DD).

Total Ionization Dose Effects (TID)

TID results in the change or loss of parameters of electronic piece such as leakage current, threshold voltage shift, and in semiconductors, it produces the electron hole -pairs in the dielectric layers and the secondary particles

[13]. Active and high energy ions can damage materials by breaking or rearranging atomic bands. In general, electrical and electronic components, a lot of materials such as insulation and dielectric capacitors, insulators cables and circuit board materials, may lead to the reduction of the insulation or electrical leakage after enough exposure to total ionization radiation. Similarly conductor materials, such as the resistance of metal plates, the TID radiation exposure can be changed or charged by exposing TID radiation. Nowadays, the most effective element influenced by the nuclear radiation in electronic circuits and telecommunications is a MOSFET transistor. Gate oxide, an ideal insulator is made of a silicon dioxide. There is a problem when the piece is exposed to radiation. First the gate oxide is ionized by absorbed dose, and free electrons and holes movement are influenced by electric field caused by the gate voltage, in oxide. The electrons are safe if they go out the oxide and disappear. But except electrons, a small fraction of holes trap in the gate and change the parameter [3]. By creating energy (TID), the piece will be turned on even if the voltage is not controlled. As a result, gate doesn't control the drain-source transistor the piece will remain on. PMOS transistors and operate similar but in a different direction. When the radiated ray traps enough positive charge in the gate oxide, the transistor remains off [14]. Moreover charge storage influences, the characteristic of voltage-current transistors (Figures 8 and 9). Threshold voltage moves by exposing transistors to radiations and enough dose storage (figure 8) and it increases the leakage current, so power consumption increases and the chip is defected (**Figure 9**(b)). Moreover, the on and off position of the transistor cannot be controlled anymore [6].

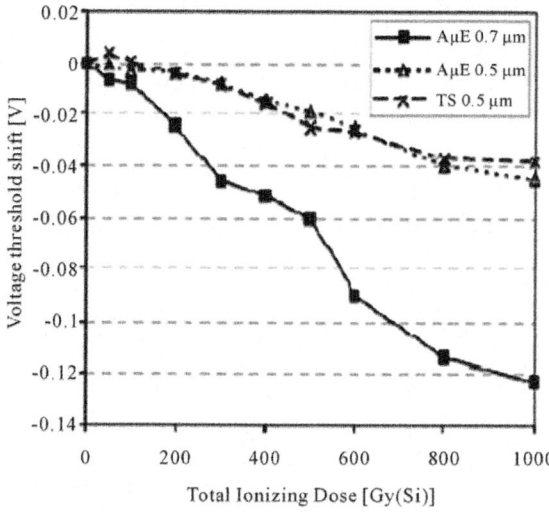

Figure 8. Threshold voltage shift.

One of the methods of decreasing these effects is to reduce the gate oxide layer. So thinner gate oxide, results in trapping less positive charge and CMOS transistors become more resistant against radiation. By decreasing transistors sizes, they also receive lower dose and have less parameter changes. TID effects will be decreased by putting transistors inside the covered chamber and shielding electronic circuits [4].

The shield effect on the charged particles in the nuclear radiation is using MCNP and Fluka software shown in figure 10 [7,8].

Displacement Damage Effect (DD)

The first known resources causing DD are protons and radio isotopes producing neutrons. When high energy proton and neutron particles influence inside a semiconductor crystal network like silicon, there are several mechanisms to move atoms. Elastic scattering is an example. Some of the particles can transfer energy to the silicon core in this phenomenon. If enough energy transfers (almost 25 ev) the core exits of its location [6]. Released silicon atom, can lose its energy by ionization changing location by the other atoms. For example in a JFET transistor, it is exposed to neutrons or protons and new recombination's centers are created. So electron recombination with holes in the base will probably be increased and by increasing the flux of neutrons or protons, it reduces transistor gain [12]. Some charges scattered in displacement damage, reduces the transistor conductivity and increase, the noise level. Displacement damage effects cause the tunneling phenomenon, increasing the current, bias change and the electrical field change in the semiconductor segment.

Single Event Effects (SEE)

The energy of ion loses in its path within the semiconductor device, creates the intense current with a maximum peak of several hundred micro-amps in a short time (less than 9 ns) at the electrical nodes.

This unwanted current leads to stable disturbances. Also single event effects transient disturbance damage cause instantaneous changes in the analog signals [11]. SEE effects are stable disorder which result to the moment changes in digital signals and it just damages the data [12]. The dielectric charge is the other effect. The radiation sources cause this, so electrons enter the domestic environment of spacecraft nuclear equipment and it results the accumulation of charge in electric insulators, cable sheathing, electronic boards and electrical fittings.

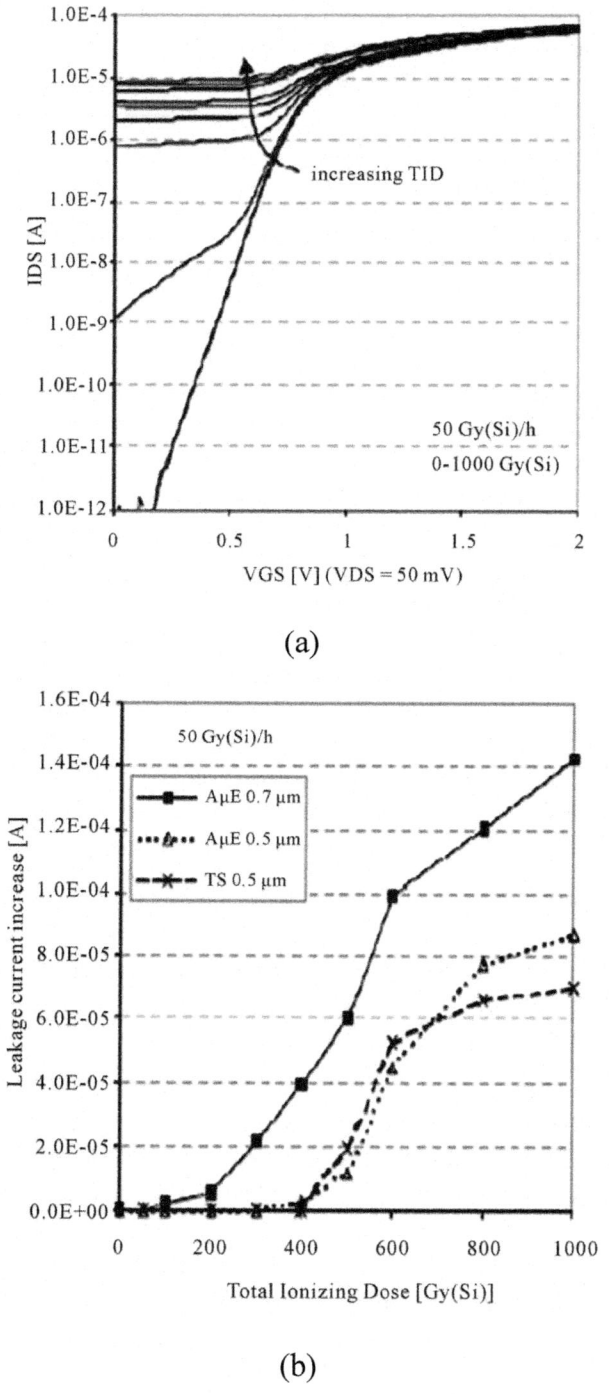

(a)

(b)

Figure 9: (a) The curve of current-voltage changes; (b) leakage current creation.

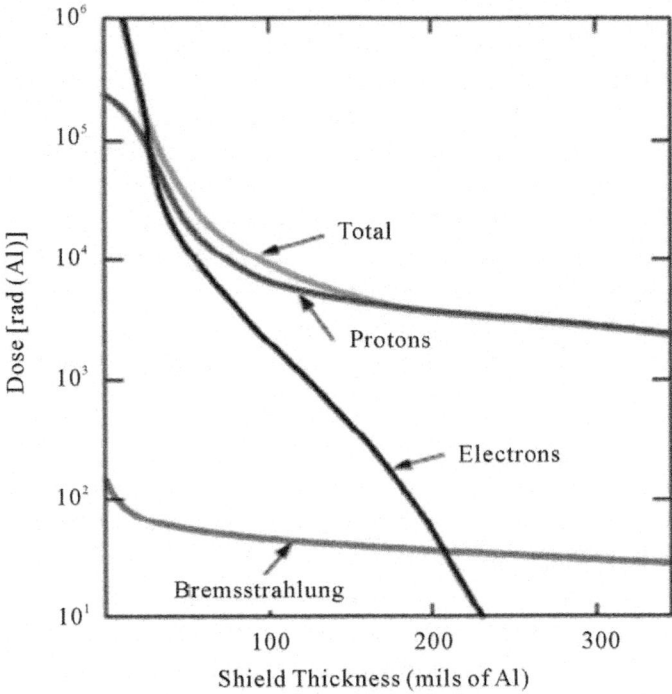

Figure 10: The shield effect on the TID.

Charging and discharging electrons also causes breakdown voltage in components. in addition to shielding for countering the single event effects it is necessary to have the various circuit techniques to reduce single event effects as well [5]. Smalling circuit dimensions and reducing the transistors size leads to reducing charges which single event effect creates in circuit elements. In addition, circuit techniques, "vote logic" can be used. It is used to correct potential errors in latches. The latch cannot change a bit position by itself in this technique, but some similar latches are required. This situation will only be changed when the majority latches want to exit. So the other latches identify and eliminate a lonely latch error. Other circuit techniques are used, such as "error detection and correction" in addition to shield and reduction of the transistors sizes in order to reduce single event effects in logic circuits [4].

CONCLUSIONS

According to calculations, simulations and experimental work done and the results of the investigation it was determined that nuclear radiations have effects, such as the creation of defects in segment, and behavioral characteristic change and creating leakage currents and noise. Total ionization dose rate effects

(TID) and single event effects (SEE) mostly limit performance of electronic equipment and telecommunications. There are some methods to counter these effects such as shielding, reducing the transistor sizes, circuit dimensions and various circuit techniques. Also in laboratory, the effects of neutron irradiation on the electronic properties of silicon were investigated and it has been shown that a high energy particle such as neutrons has a significant effect on electronic properties of semiconductor devices. The influence was displayed as of crystal defects decreasing the conductance and increasing resistance. The thermal recovery operation was performed to repair damages and it was observed that the samples which were too near the sources of radiation or under a long time radiation can be recovered with the higher temperature (750°C).

REFERENCES

1. L. W. Townsend, J. L. Shinn and J. W. Wilson, "Interplanetary Crew Exposure Estimates for the August 1972 and October 1989 Solar Particle Events," Radiation Research, Vol. 126, No. 1, 1991, pp. 108-110. doi;10.2307/3578178

2. F. A. Cucinotta, W. Schimmerling, J. W. Wilson, L. E. Petersen, G. D. Badhwar, P. B. Saganti and J. F. Dicello, "Space Radiation Cancer Risk Projections for Exploration Missions: Uncertainty Reduction and Mitigation," DIANE Publishing, Darby, 2001.

3. T. Liu, "Total Ionization Dose Effects and Single-Event Effects Studies of a 0.25 μm Silicon-on-Sapphire CMOS Technology," 9th European Conference on Radiation and Its Effects on Components and Systems, Deauville, 10-14 September 2007, pp. 1-5.

4. E. N. Parker, "Shielding Astronauts from Cosmic Rays," Space Weather, Vol. 3, 2005, p. S08004.

5. A. H. Johnston, "Radiation Damage of Electronic and Optoelectronic Devices in Space," 4th International Workshop on Radiation Effects on Semiconductor Devices for Space Application, Tsukuba, 11-13 October 2000.

6. G. C. Messenger and M. S. Ash, "The Effects of Radiation on Electronic Systems," Van Nostrand Reinhold, New York, 1992.

7. A. Ferrari, et al., "Fluka: A Multi-Particle Transport Code," CERN, Geneva, 2005.

8. X-5 Monte Carlo Team, "MCNP-A General Monte Carlo N-Particle Transport Code," Los Alamos National Laboratory, Los Alamos, 2003.

9. T. P. Ma and P. V. Dressendorfer, "Ionizing Radiation Effects in MOS Devices and Circuits," John Wiley and Sons, New York, 1989.

10. S. Duzellier, "Radiation Effects on Electronic Devices in Space," Aerospace Science and Technology, Vol. 9, No. 1, 2005, pp. 93-99. doi;10.1016/j.ast.2004.08.006

11. G. C. Messenger and M. S. Ash, "Single Event Phenomena," Kluwer Academic Publishers, New York, 1997. doi;10.1007/978-1-4615-6043-2

12. G. Barbottin and A. Vapaille, "Instabilities in Silicon Devices," Elsevier, Berlin, 1999.

13. H. J. Barnaby, M. Mclain and I. S. Esqueda, "Total-IonIzing-Dose Effects on Isolation Oxides in Modern CMOS Technologies," Nuclear Instruments and Methods in Physics Research B, Vol. 261, No. 1-2, 2007, pp 1142-1145. doi;10.1016/j.nimb.2007.03.109

14. H. J. Barnaby, "Total-Ionizing-Dose Effects in Modern CMOS Technologies," IEEE Transactions on Nuclear Science, Vol. 53, No. 6, 2006, pp. 3103-3121.doi;10.1109/TNS.2006.885952

Chapter 3

MODELING AND SIMULATION FOR HIGH ENERGY SUB-NUCLEAR INTERACTIONS USING EVOLUTIONARY COMPUTATION TECHNIQUE

Mahmoud Y. El-Bakry[1], El-Sayed A. El-Dahshan[2,3], Amr Radi[3,4], Mohamed Tantawy[1], Moaaz A. Moussa[1,5]

[1]Department of Physics, Faculty of Sciences, Ain Shams University, Cairo, Egypt

[2]Egyptian E-Learning University, Giza, Egypt

[3]Department of Physics, Faculty of Education, Ain Shams University, Cairo, Egypt

[4]The British University in Egypt (BUE), Cairo, Egypt

[5]Buraydah Colleges, East Qassim University, Buraydah, KSA

ABSTRACT

High energy sub-nuclear interactions are a good tool to dive deeply in the core of the particles to recognize their structures and the forces governed. The current article focuses on using one of the evolutionary computation techniques, the so-called genetic programming (GP), to model the hadron nucleus (h-A) interactions through discovering functions. In this article, GP is used to simulate the rapidity distribution $\left(\frac{1}{N}\frac{dN}{dY}\right)$ **of total charged, positive and negative** pions for p^--Ar and p^--Xe interactions at 200 GeV/c and charged particles for p-pb collision at 5.02 TeV. We have done so many runs to select the best runs of the GP program and finally obtained the rapidity distribution $\left(\frac{1}{N}\frac{dN}{dY}\right)$ **as a function of the lab momentum** $\left(P_{Lab}\right)$, mass number (A) and the number of particles per unit solid angle (Y). In all cases studied, we compared our seven discovered functions produced by GP technique with the corresponding experimental data and the excellent matching was so clear.

INTRODUCTION

Evolutionary computation refers to a class of algorithms that utilize simulated evolution to some degree as a means to solve a variety of problems, from numerical optimization to symbolic logic. By simulated evolution, we mean that the algorithms have the ability to evolve a population of potential solutions such that weaker solutions are removed and replaced with incrementally stronger (better) solutions. In other words, the algorithms follow the principle of natural selection. Each of the algorithms has some amount of biological plausibility, and is based on evolution or the simulation of natural systems [1] - [8] .

In 1990s, John Koza [9] [10] introduced the subfield called Genetic Programming. This is considered a subfield because it fundamentally relies on the core genetic algorithm created by Holland [11] , and differs in the underlying representation of the solutions to be evolved. Instead of using bit-strings (as with genetic algorithms) or real-values (as is the case for evolutionary programming [1] - [7] or evolutionary strategies [1] - [7] , genetic programming relies on S-expressions (program trees) as the encoding scheme.

Hadron-nucleus (h-A) interactions have been considered as a corner stone in high energy physics because of its theoretical and practical interesting features and also it is an intermediate state between hadron-hadron (h-h) and nucleus-nucleus (N-N) interactions. So, there are a lot of models that concern the study of the hadron structure [12] - [16] and the interactions between hadrons and nuclei such as the three-fireball model [17] , quark models [18] - [20] , fragmentation model [21] - [23] and many more.

In our previous works [24] - [27] , our group studied the applications of artificial intelligence and the evolutionary computation techniques such as neural network, adaptive fuzzy inference system, genetic programming, genetic algorithm, hybrid technique model and many others to solve many complex (nonlinear) problems in high energy physics and showed best fitting with the corresponding experimental data in comparison with the conventional techniques.

The study of hadron-nucleus interaction at high and ultrahigh energy has been a subject of great interest to high energy physicist because the nuclei provide a number of unique physics opportunities which are not available in elementary particle collisions [28] .

In this article, Genetic programming (GP) model has been used to discover a function that computes the rapidity distribution of created (total charged, positive and negative) pions for p^--Ar and p^--Xe collisions at 200 GeV/c [29] - [31] and charged particles for p-pb collision at 5.02 TeV [32] . The seven discovered functions produced by GP model show an excellent matching when they have been compared to the corresponding experimental data [29] - [32] . This article is organized as follows; Section 2 gives the definition and the outlines to the basics of the GP technique. Section 3 reviews the implementation of GP. Finally, the results and conclusions are provided in Sections 4 and 5 respectively.

GP OUTLINES

GP is defined as the biologically-inspired evolution of computer programs that solve a predefined task. For this reason, GP is nothing more than a genetic algorithm applied to the problem program evolution. Early GP systems utilized LISP (the original functional programming language, 1958) S-expressions (as shown in Figure 1), but more recently, linear GP systems have been used to evolve instruction sequences to solve user-defined programming tasks [2] .

Since genetic programming manipulates programs by applying genetic operators (reproduction, crossover and mutation), a programming language should permit a computer program to be manipulated as data and the newly created data to be executed as a program. For these reasons, LISP was chosen as the main language for GP.

Evolving complete programs with GP is computationally very expensive, and the results have been limited, but GP does have a place in the evolution of program fragments. For example, the evolution of individual functions that have very specific inputs and outputs and whose behavior can be easily defined for fitness evaluation by GP. To evolve a function, the desired output must be easily measurable in order to understand the fitness landscape of the function in order to incrementally evolve it [1] - [7] .

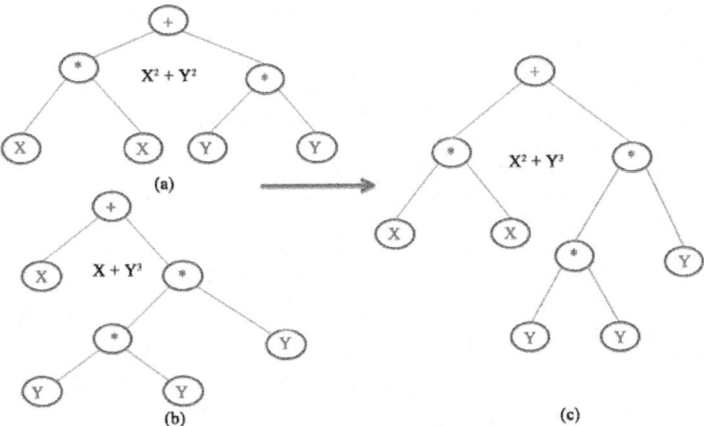

Figure 1: Using the crossover operator to create a new S-expression.

Consider the example shown in Figure 1. The population consists of two members, A and B. Using the crossover operator, a portion of A is grafted onto B; resulting in a new expression C. GP also utilizes the mutation operator as a way of extending the population to the search space.

GP IMPLEMENTATION

The GP uses the same fundamental flow as the traditional genetic algorithm. The population of potential solutions is initialized randomly and then their fitness computed (through a simulation of executed instructions with the stack). Selection of members that can propagate into the next generation can then occur through fitness proportionate selection. With this method, the higher fit the individual, the higher the probability that they will be selected for recombination in the next generation. Evolutionary algorithms borrow concepts from Darwinian natural selection as a means to evolve solutions to problems, choosing from more fit individuals to propagate to future generations [1] - [7] .

The chromosome, or program to be evolved, is made up of genes, or individual instructions. The chromosome can also be of different lengths, assigned at creation, and then inherited during the evolution.

All methods of evolutionary computation (and then GP) work as follows: create a population of individuals, evaluate their fitness, generate a new population by applying genetic operators (Cross-over, mutation and reproduction), and repeat this process a number of times as shown in Figure 2.

RESULTS AND DISCUSSION

We have performed the GP modeling of the inclusive reaction,

$$p^- + Ar, Xe, pb \rightarrow \pi^\pm + X$$

$$p + pb \rightarrow \text{charged particles}$$

(1)

using the experimental data [29] - [32] at 100 and 200 GeV/c and have done so many runs to select the best runs of the GP program, the first runs are for simulating the rapidity distribution $\frac{1}{N}\frac{dN}{dY}$ of negative pions for p^--Au, Ag, Mg collisions at 100 GeV/c. They were configured to have the lab momentum (P_{Lab}), mass number (A) and the number of particles per unit solid angle (Y) as inputs and the output is the corresponding rapidity distribution $\left(\frac{1}{N}\frac{dN}{dY}\right)$ of negative pions at the given momentum as shown in Figure 3.

The second ones are for simulating the rapidity distribution $\left(\frac{1}{N}\frac{dN}{dY}\right)$ of positive pions for p^--Au, Ag, Mg collisions at 100 GeV/c as an output and the inputs are the same as in Figure 3.

The third ones are for simulating the rapidity distribution $\left(\frac{1}{N}\frac{dN}{dY}\right)$ of positive pions for p^--Ar, p^--Xe collisions at 200 GeV/c as an output and the inputs are the same as in Figure 3.

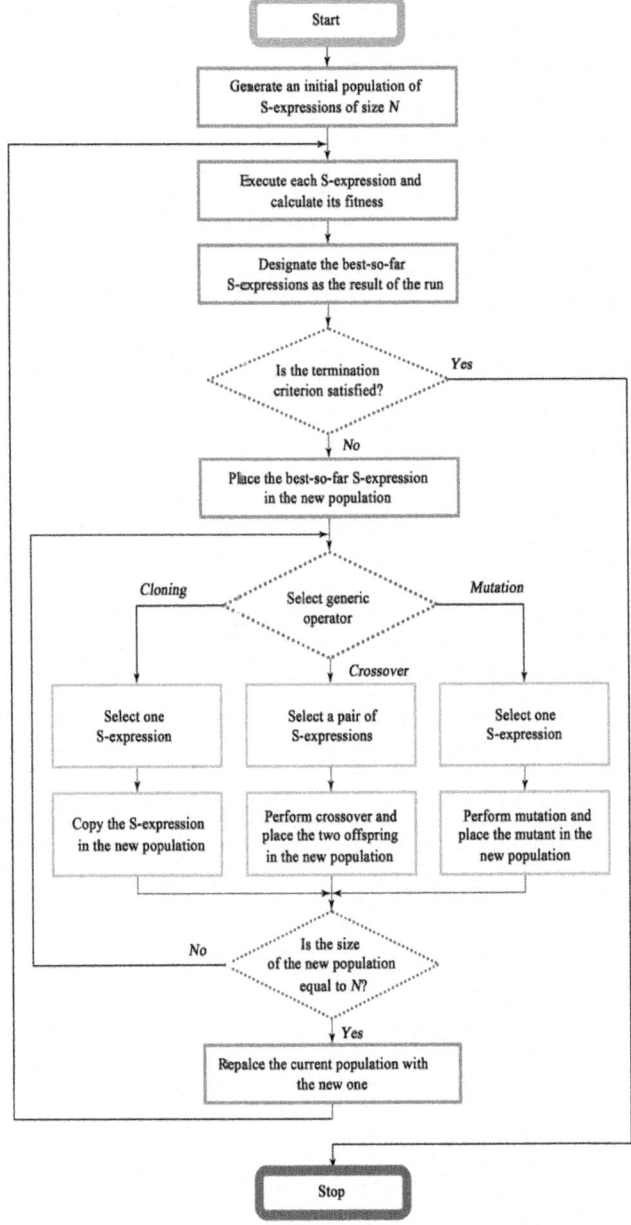

Figure 2: Flowchart for GP.

Inputs   Output

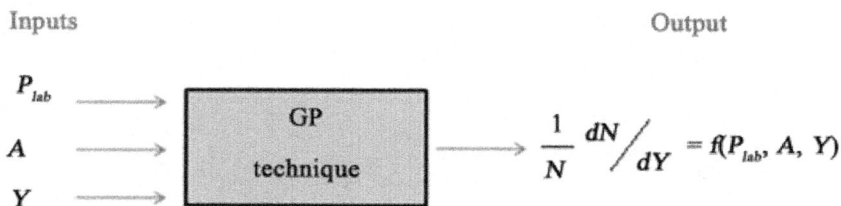

Figure 3: A block diagram of the GP technique.

The fourth ones are for simulating the rapidity distribution $\left(\frac{1}{N}\frac{dN}{dY}\right)$ of negative pions for p^--Ar, p^--Xe collisions at 200 GeV/c as an output and the inputs are the same as in Figure 3.

The fifth ones are for simulating the rapidity distribution $\left(\frac{1}{N}\frac{dN}{dY}\right)$ of charged pions for p^--Ar, p^--Xe collisions at 200 GeV/c as an output and the inputs are the same as in Figure 3.

The last ones are for simulating the rapidity distribution $\left(\frac{1}{N}\frac{dN}{dY}\right)$ of charged particles for p-pb collisions at 5.02 TeV as an output and the inputs are the same as in Figure 3.

According to all the runs, we have obtained the corresponding tree and their equivalent discovered functions Equations (2)-(7) generated for the rapidity distribution $\left(\frac{1}{N}\frac{dN}{dY}\right)$ of total charged, positive and negative pions for p^--Ar collisions at 100, 200 GeV/c and 5.02 TeV.

The output, the rapidity distribution$\left(\frac{1}{N}\frac{dN}{dY}\right)$, as a function of the inputs (P_{Lab}, A, Y) is given as follows:

For negative pions at 100 GeV/c,

$$\frac{1}{N}\frac{dN}{dY} = \left[\sin^2(F) - \sin\left(e^E\right)\right]$$

(2)

For positive pions at 100 GeV/c,

$$\frac{1}{N}\frac{dN}{dY} = e^{\left[0.8613\sin(X_1)/R\right]+E}$$

(3)

For positive pions at 200 GeV/c,

$$\frac{1}{N}\frac{dN}{dY} = \sin\left(H + C\right) + e^{J}$$

(4)

For negative pions at 200 GeV/c,

$$\frac{1}{N}\frac{dN}{dY} = \sin\left\{X_1\left[\sin\left(\cos^2\left(10\right)\right)\right]\right\} + J$$

(5)

For charged pions at 200 GeV/c,

$$\frac{1}{N}\frac{dN}{dY} = \sin^4\left(0.62932X_1\right) + b$$

(6)

For charged particles at 5.02 TeV,

$$\frac{1}{N}\frac{dN}{dY} = E + U$$

(7)

For more details about F, E,H, U, etc., see Appendices 1-6.

The comparison between the pions rapidity distribution $\left(\frac{1}{N}\frac{dN}{dY}\right)$ computed by employing our discovered functions Equations (2)-(7) and the corresponding experimental data [29] - [31] are represented in Figure 4 for negative pions for p^--Au, Ag, Mg collisions at 100 GeV/c, Figure 5 for positive pions for p^--Au, Ag, Mg

collisions at 100 GeV/c, Figure 6 for positive pions for p^--Ar (the GP model cannot describe the data when the axial value is near 0 because the noisy behavior of data around the axial), p^--Xe collisions at 200 GeV/c. Figure 7 for negative pions for p^--Ar, p^--Xe collisions at 200 GeV/c, Figure 8 for charged pions for p^--Ar, p^--Xe collisions at 200 GeV/c, Figure 9 for charged particles for p-pb collisions at 5.02 TeV.

In order to generate the GP model we have implemented the GP steps (fitness evaluation, reproduction, crossover and mutation) that were mentioned in Section 3. Our six discovered functions are generated using the obtained control GP parameters, which are shown in Table 1.

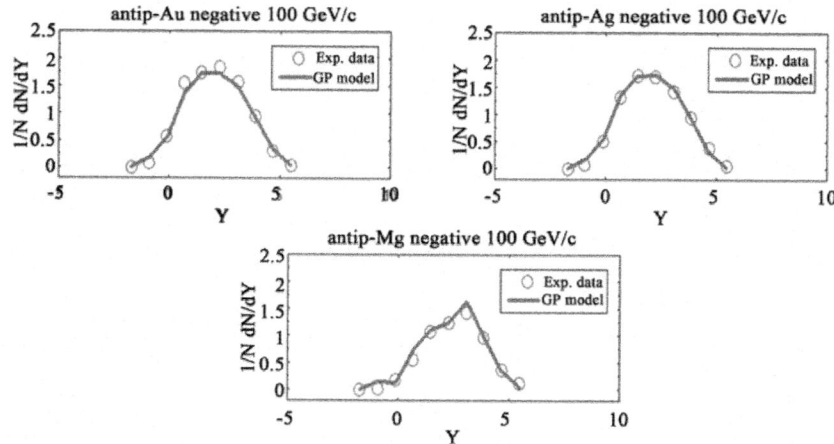

Figure 4: The discovered rapidity distribution of negative pions $\left[\frac{1}{N}\frac{dN}{dY}\right]$ for antip-Au, antip-Ag, antip-Mg interaction at 100 GeV/c: (—) GP model, (o) experimental data.

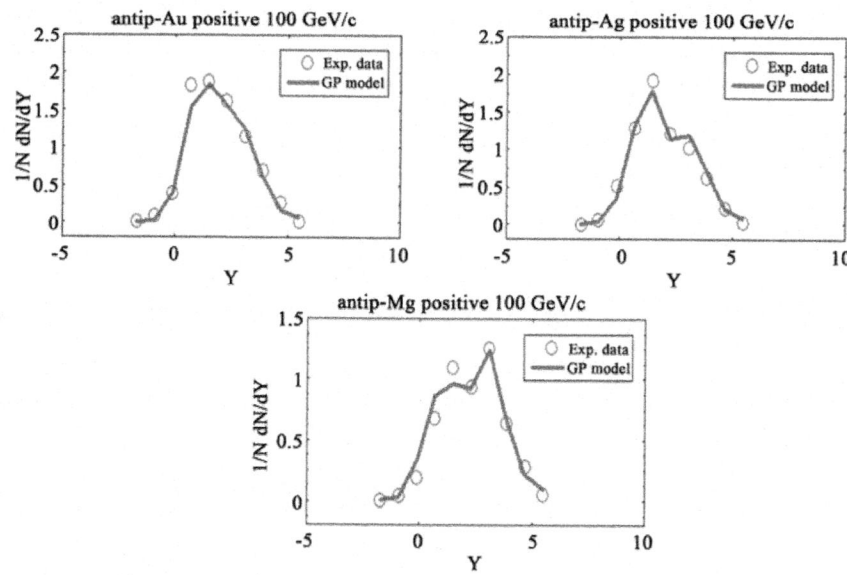

Figure 5: The obtained rapidity distribution of positive pions $\left[\frac{1}{N}\frac{dN}{dY}\right]$ for antip-Au, antip-Ag, antip-Mg interaction at 100 GeV/c: (—) GP model, (o) experimental data.

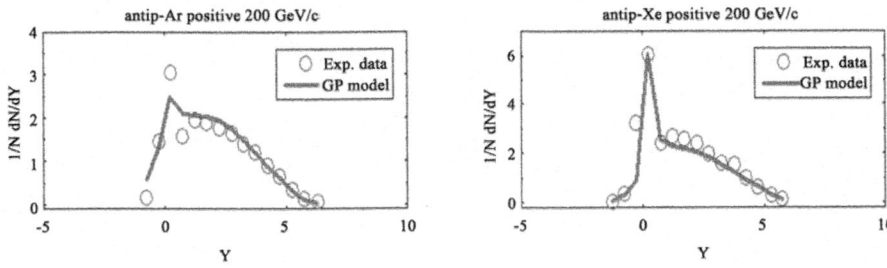

Figure 6: The discovered rapidity distribution of negative pions $\left(\frac{1}{N}\frac{dN}{dY}\right)$ for antip-Au, antip-Ag, antip-Mg interaction at 100 GeV/c: (—) GP model, (o) experimental data.

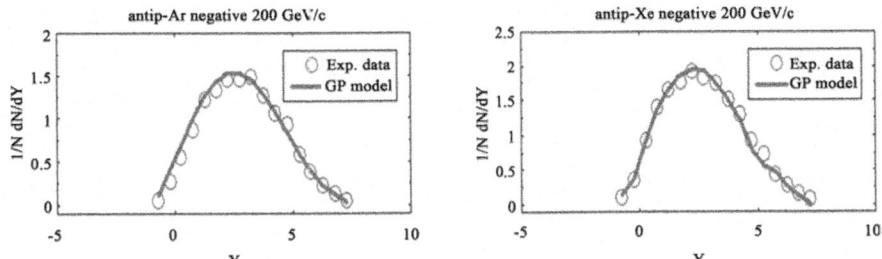

Figure 7: GP-simulated for rapidity distribution of negative pions $\left(\frac{1}{N}\frac{dN}{dY}\right)$ for antip-Ar, antip-Xe at 200 GeV/c: (—) GP model, (o) experimental data.

Table 1: Optimal parameters controlling GP program

GP parameters	Total charged pions/particles	Positive pions	Negative pions
Individuals	1000	1000	1000
Generations	50	50	50
Function set	+, −, *, /, ln, log, sin, cos, tan, sqrt, exp, power	+, −, *, /, ln, log, sin, cos, tan, sqrt, exp, power	+, −, *, /, ln, log, sin, cos, tan, sqrt, exp, power
Terminal set	P_{Lab}, A and Y	P_{Lab}, A and Y	P_{Lab}, A and Y
Fitness function	MSE	MSE	MSE
Selection	Tournament	Tournament	Tournament
Fitness at 5.02 TeV	0.9673	-	-
Mutation rate	0.01	-	-
Crossover rate	0.9	-	-
Fitness at 200 GeV/c	4.9503	6.8111	1.8313
Mutation rate	0.9	0.01	0.01
Crossover rate	0.01	0.9	0.9
Fitness at 100 GeV/c	-	2.178635	1.5753
Mutation rate	-	0.01	0.9
Crossover rate	-	0.9	0.01

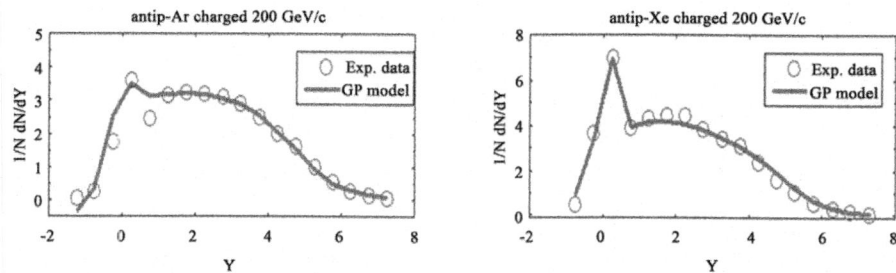

Figure 8. Simulated results for rapidity distribution of chared pions $\left(\frac{1}{N}\frac{dN}{dY}\right)$ for antip-Ar, antip-Xe at 200 GeV/c: (—) GP model, (o) experimental data.

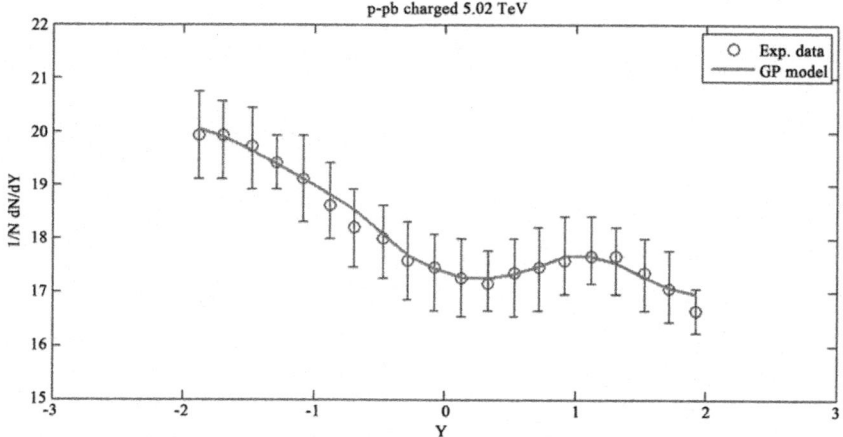

Figure 9: The discovered rapidity distribution of charged particles for p-Pb interaction at 5.02 TeV: (—) GP model, (o) experimental data.

The statistical error criterion of mean square error (MSE) was used to measure the deviation between the experimental (actual) and simulated values. The statistical parameter MSE has been used in this work as a performance metric [33] to compare the GP simulation with the actual observations (experimental data) and these were evaluated by using Matlab program. The smaller the values of MSE the closer the simulated values to the experimental ones. Our obtained MSE values for the seven discovered functions are given in Table 1, which show also that the performance of the GP model is clearly suitable.

CONCLUSIONS

GP model has been shown to be a vital method for modeling the h-A interactions. The current article presents an efficient approach for computing the rapidity distribution $\left(\frac{1}{N}\frac{dN}{dY}\right)$ of charged, positive and negative pions for p^--Ar and p^--Xe collisions at 200 GeV/c and charged particles for p-pb collisions at 5.02 TeV through the obtained discovered functions. All the discovered functions show a clear and excellent match to the experimental data.

The interaction of hadrons with atomic nuclei at high and ultrahigh energies is an issue of great importance since in-depth studies about it provide the necessary information on properties which cannot be examined by analyzing only hadron-hadron interactions.

Finally, the present work has proved that the GP approach can be employed effectively to model the h-A interactions at the given energy.

REFERENCES

1.   Jones, M.T. (2008) Artificial Intelligence: A Systems Approach Infinity. Science Press LLC, Hingham.

2.   Banzhaf, W., et al. (1998) Genetic Programming: An Introduction: On the Automatic Evolution of Computer Programs and Its Applications. Morgan Kaufmann, Burlington.

3.   Higuchi, T., Liu, Y. and Yao, X. (2006) Evolvable Hardware. Genetic and Evolutionary Computation. http://dx.doi.org/10.1007/0-387-31238-2

4.   EvoNews Professor Hans-Paul Schwefel Talks to EvoNews (1999) Available Online. http://evonet.lri.fr/evoweb/news_events/news_features/

5.   Fogel, L.J., Owens, A.J. and Walsh, M.J. (1966) Artificial Intelligence through Simulated Evolution. Wiley, New York.

6.   Levenick, J.R (1991) Inserting Introns Improves Genetic Algorithm Success Rate: Taking a Cue from Biology. Proceedings on the 4th International Conference on Genetic Algorithms.

7.   Rechenberg I (1965) Cybernetic Solution Path of an Experimental Problem Technical Report Library Translation No. 1122. Royal Aircraft Establishment, Farnborough.

8.   Zheng, S.J., Zhang, N., Xia, Y.J. and Wang, H.T. (2014) Research on Non-Uniform Strain Profile Reconstruction along Fiber Bragg Grating

via Genetic Programming Algorithm and Interrelated Experimental Verification. Optics Communications, 315, 338-346.

9.   Koza, J.R. (1992) Genetic Programming: On the Programming of Computers by Means of Natural Selection. The MIT Press, Cambridge.

10.   Koza, J.R. (1990) Genetic Programming: A Paradigm for Genetically Breeding Populations of Computer Programs to Solve Problems. Stanford University Computer Science Department Technical Report STAN-CS-90-1314.

11.   Holland, J.H. (1975) Adaptation in Natural and Artificial Systems. University of Michigan Press, Ann Arbor.

12.   Tantawy, M., El-Mashad, M. and El-Bakry, M.Y. (1998) Multiparticle Production Process in High Energy Nucleus-Nucleus Collisions. Indian Journal of Physics, 72A, 73-82.

13.   Moussa, M.A., El-Bakry, M.Y., Radi, A., El-Dahshan, E.-S.A., Habashy, D.M. and Abbas, E.G. (2012) Topological Cross Sections and Multiplicity Distributions for and Interactions at High Energies. International Journal of Scientific and Engineering Research, 3.

14.   Fermi, E. (1950) High Energy Nuclear Events. Progress of Theoretical Physics, 5, 570-583. http://dx.doi.org/10.1143/ptp/5.4.570

15.   Fermi, E. (1951) Angular Distribution of the Pions Produced in High Energy Nuclear Collisions. Physical Review, 81, 683-687. http://dx.doi.org/10.1103/PhysRev.81.683

16.   Ranft, J. (1970) Secondary Particle Production According to the Thermodynamical Model and New Experimental Data. Physics Letters B, 31, 529-532. http://dx.doi.org/10.1016/0370-2693(70)90082-1

17.   Xu, C., Chao, W.-Q., Meng, T.-C. and Huang, C.-S. (1986) Statistical Approach to Nondiffractive Hadron-Hadron Collisions: Multiplicity Distributions and Correlations in Different Rapidity Intervals. Physical Review D, 33, 1287-1299. http://dx.doi.org/10.1103/PhysRevD.33.1287

18.   Nambu, Y. (1976) The Confinement of Quarks. Scientific American, 235, 48-61. http://dx.doi.org/10.1038/scientificamerican1176-48

19.   Gyulassy, M. (1985) Introduction to QCD Thermodynamics and the Quark-Gluon Plasma. Progress in Particle and Nuclear Physics, 15, 403-442. http://dx.doi.org/10.1016/0146-6410(85)90076-6

20.   Kisslinger, L.S. (1985) Nuclear Physics and Quark/Gluon QCD. Nuclear Physics A, 446, 479-488. http://dx.doi.org/10.1016/0375-9474(85)90624-4

21.   Jacob, M. and Slansky, R. (1972) Nova Model of Inclusive Reactions. Physical Review D, 5, 1847-1870. http://dx.doi.org/10.1103/PhysRevD.5.1847

22.   Hwa, R.C. (1970) Bootstrap Model for Diffractive Processes: Complementarity of the Yang and Regge Models. Physical Review D, 1, 1790-1809. http://dx.doi.org/10.1103/PhysRevD.1.1790

23.   Hwa, R.C. (1971) Multiplicity Distribution and Single-Particle Spectrum in the Diffractive Model. Physical Review Letters, 26, 1143-1147. http://dx.doi.org/10.1103/PhysRevLett.26.1143

24.   EL-Bakry, S.Y., El-Dahshan, E.-S. and EL-Bakry, M.Y. (2011) Total Cross Section Prediction of the Collisions of Positrons and Electrons with Alkali Atoms Using Gradient Tree Boosting. Indian Journal of Physics, 85, 1405-1415.

25.   El-Bakry, M.Y. (2003) Feed Forward Neural Networks Modeling for K-P Interactions. Chaos, Solitons and Fractals, 18, 995-1000. http://dx.doi.org/10.1016/S0960-0779(03)00068-7

26.   El-Bakry, M.Y. (2004) A Study of K-P Interaction at High Energy Using Adaptive Fuzzy Inference System Interactions. International Journal of Modern Physics C, 15, 1013-1020.http://dx.doi.org/10.1142/S0129183104006467

27.   El-Bakry, M.Y., El-Dahshan, E., Radi, A., Tantawy, M. and Moussa, M.A. (2013) A Genetic Programming for Modeling Hadron-Nucleus Interactions at 200 GeV/c. International Journal of Scientific and Engineering Research, 4, 7.

28.   Ghosh, D. (1983) International Conference on Cosmic Ray 08.

29.   De Marzo, C., De Palma, M., Distante, A., et al. (1982) Multiparticle Production on Hydrogen, Argon, and Xenon Targets in a Streamer Chamber by 200-GeV/c Proton and Antiproton Beams. Physical Review D, 26, 1019-1035. http://dx.doi.org/10.1103/PhysRevD.26.1019

30.   Arneodo, M., Arvidson, A., Aubert, J.J., et al. (1987) Comparison of Multiplicity Distributions to the Negative Binomial Distribution in Muon-Proton Scattering. Zeitschrift für Physik C Particles and Fields, 35, 335-345. http://dx.doi.org/10.1007/BF01570769

31.   Kittle, W. (1973) Combining Inclusive and Exclusive Data Analyses—What Have We Learned So Far? Journal of Physics A: Mathematical, Nuclear and General, 6, 733.

32.   Abelev, B., et al., ALICE Collaboration (2013) Pseudorapidity Density of Charged Particles in p + Pb Collisions at s N N=5.02 TeV. Physical

Review Letters, 110, Article ID: 032301. http://dx.doi.org/10.1103/PhysRevLett.110.032301

33. Hong, W.-C. (2008) Rainfall Forecasting by Technological Machine Learning Models. Applied Mathematics and Computation, 200, 41-57. http://dx.doi.org/10.1016/j.amc.2007.10.046

Appendices

1. Rapidity Distribution of Negative Pions for p⁻Au, p⁻Ag and p⁻Mg Interaction at 100 GeV/c

$$A = \sin^2\left(10 - \frac{X_1 + X_3}{1.2414}\right), \quad B = A\left\{\log_{10}\left[\cos^2\left(X_1 + e^{X_3}\right)\right]\right\}, \quad C = \sin^2\left(10 - \frac{X_1 + 0.60169}{1.2414}\right),$$

$$D = \log_{10}\left[\cos\left(C + \sin\left(e^B\right)\right)\right], \quad E = \frac{0.17D}{X_1 e^{\sin(X_3)}}, \quad F = 10 - \left[\frac{X_1 + 0.60169}{1.2414}\right]$$

2. Rapidity Distribution of Positive Pions for p⁻Au, p⁻Ag and p⁻Mg Interaction at 100 GeV/c

$$A = X_2 - X_1 - e^{X_1} + X_3 - 970.8738, \quad B = X_2 - 0.12013 - \tan(A), \quad C = (10/B)^{\sin(X_3)} - \cos(X_3)$$

$$D = \cos^2(0.187454 + C), \quad E = \sin^2(0.550985 X_1) - D,$$

$$F = \cos\left(\log_{10}\left(\tan(X_3)\right) + \left(\tan\left(e^{X_3}\right) - \cos(X_3)\right)\right), \quad G = \tan(X_3) + X_3 X_2 - \left(X_1 + 10/X_2\right)$$

$$H = \sin\left[(G)^{10}/F\right], \quad I = \frac{-2.91765}{10/X_3}, \quad J = \frac{(10/X_3)^I - 0.82908 + H}{X_3},$$

$$K = \log_{10}\left(\tan(X_3)\right) + \cos(X_3) - \tan\left(e^{X_3}\right), \quad L = X_2^{X_3} + X_3 X_2 - 10,$$

$$M = \left[\frac{X_1 + 10}{X_2}\right]^{\sin(X_3)} - 0.82908, \quad N = \sin\left[\frac{-3.64929}{10/X_3}\right], \quad O = \cos\left[\sin\left(10/X_3\right)^N + m\right],$$

$$P = \sin\left(\frac{\cos(K)}{L^O}\right) + J^I - 0.8298, \quad Q = \sin(0.176 X_1) - \cos^3(P), \quad R = \tan\left(e^Q\right)$$

3. Rapidity Distribution of Positive Pions for p⁻Xe and p⁻Ar Interaction at 200 GeV/c

$$A = \frac{1.228815}{[X_1/0.831998]}, \quad B = \frac{e^{\sin(A)}}{X_1/0.724939}, \quad C = X_1 \cos\left[0.984928 \sin\left(e^{\sin(B)}\right)\right]$$

$$D = \frac{\left[10/\cos\left(\sin\left(X_1\right)\right)\right]}{\left[56.2934 - \log_{10}\left(X_1\right)\right]}, \quad E = \sin\left(X_1\right) - \cos^2\left[\left(D-10\right)\log_{10}\left(X_1\right)\right],$$

$$G = 0.984928\sin\left(e^{\sin(A)}\right), \quad H = 0.984928\sin^2\left(G+E\right), \quad I = 0.942092\cos\left(X_2\right), \quad J = \frac{e^I}{\left(X_1/0.20605\right)}$$

4. Rapidity Distribution of Negative Pions for p⁻Ar and p⁻Xe Interaction at 200 GeV/c

$$A = X_3 \Big/ \left[10X_1 / 10e^{X_1 + \cos(X_2)}\right], \quad B = \sin\left[X_1\left(0.995888 + \frac{10X_1}{0.39473}\right)\right]$$

$$C = e^{\left(e^{X_1 + 0.66922}\right)} - \cos^2\left(X_3\right) \quad D = 10\left[\frac{X_1}{e^{X_2}}\right]\Big/0.39473 \quad E = \left(C - \frac{X_3}{D}\right)\Big/B$$

$$G = \cos^2\left[\sin\left(F\right)\right] - X_3 \Big/ \left(\frac{10X_1}{0.39473}\right), \quad H = \log_{10}\left(G\right) - \left(X_3 \Big/ \frac{10X_1}{0.39473}\right), \quad I = \sin\left[e^{\sin(X_2)}\right]\Big/X_1,$$

$$J = \cos\left[\sin\left(\frac{e^{\sin(X_3)}X_1}{I}\right)\right]$$

5. Rapidity Distribution of Charged Pions for p⁻Xe and p⁻Ar Interaction at 200 GeV/c

$$A = \left\{\left[\sin\left(X_1 X_2\right)\right]^{X_3} - \frac{X_1}{0.9331}\right\}^{0.13958}, \quad B = \log_{10}\left[\frac{A^{7.6003} - \left(\frac{X_2}{X_1}\right)}{0.533419}\right], \quad C = \sin^2\left(0.62932X_1\right) + B$$

$$D = \sin\left[\sin^2\left(0.62932X_1\right) - 0.37885\right], \quad E = X_2^{0.13958} - \left[\frac{\frac{X_2}{X_1} + X_2}{X_1}\right],$$

$$F = \sin^2\left(0.62932X_1\right) + \log_{10}\left(\frac{E}{D}\right),$$

$$G = \left\{\left[\sin\left(F\right)\right]^C - 1\right\}^{0.13958}, \quad H = \left(X_2^{0.88504} + X_2\right)^{0.76003}, \quad I = \left[\frac{G^H - \left(\frac{X_2}{X_1}\right)}{0.533419}\right],$$

$$J = \sin\left[\sin^2\left(0.62932X_1\right) + \log_{10}\left(I\right)\right], \quad K = \sin\left(0.62932X_1\right) - \tan\left[\sin\left(X_2\right)\right],$$

$$L = 0.841471\cos\left(X_1\right)^K, \quad M = \left\{L - \sin\left[\cos\left(\frac{X_3}{0.88504X_2}\right) + \cos\left(X_3^2\right)\right]\right\}^{0.13958}$$

$$N = M - \frac{\frac{X_2}{X_1} + X_2}{X_1}$$

$$O = \frac{N}{\sin\left[\sin^2(0.62932X_1) + \log_{10}(0.41797)\right]}$$

$$P = \sin\left[\sin^2(0.62932X_1) + \log_{10}(O)\right]$$

$$Q = \left\{\left[\sin(X_2 X_1)\right]^{X_3} - \frac{X_1}{0.9331}\right\}^{0.13958}$$

$$R = \left\{\left[\sin(X_3)\right] + X_2^{0.88504}\right\}^{0.76003}$$

$$S = \log_{10}\left\{Q^R - \frac{X_2/X_1}{0.533419}\right\}^{0.13958}$$

$$T = \sin^2(0.62932X_1) + S$$

$$U = e^{\frac{\log_{10}(X_3) + 0.010309}{X_3 + X_2 X_3}}$$

$$V = \left[\cos(P^T)\right]^{e^U}$$

$$W = \sin\left\{10\left[\frac{X_3}{\frac{e^{X_1}}{-0.04686} + \cos(X_3^2)}\right]\right\}$$

$$X = \left\{\left[\sin(X_1 X_3)\right]^{X_3} - 1\right\}^{0.13958}$$

$$Y = \left\{\left[X_1 - \sin(X_2)\right]^{X_3} + X_2^{0.88504}\right\}^{0.76003}$$

$$Z = (VW - X_1)^{\sin(X^Y)}$$

$$a = \left[Z - \frac{(X_2/X_1) + X_2}{X_1}\right]\Big/ J$$

$$b = \sin^2(0.62932X_1) + \log_{10}(a)$$

6.   Rapidity Distribution of Charged Particles for p-pb Interaction at 5.02 TeV

$$A = (-0.05142)\frac{(10 - x_1)}{0.949882}$$

$$B = -9.64173\cos(X_1)$$

$$C = (0.786251)^B$$

$$D = C + \log_2\left[\sin(10) - \log_2(10)\right]$$

$$E = \frac{\log_2(e^D)}{22026.47}$$

$$F = \log\left(\frac{0.97842}{X_1}\Big/10\right)\left[\log_{10}(e^{X_1}) + X_1\right]$$

$$G = 1.9999$$

$$H = \cos\left[\cos(X_1)\right] - \sqrt{G}$$

$$I = \log\left[\log_{10}(0.4439)\right] - \tan\left\{\tan\left[\log(X_1) - \sqrt{X_1}\right]\right\}$$

$$J = H \tan\left(e^{\log_{10}\left(\log_{10}(I)\right)}\right), \quad K = e^{\log_{10}(X_1)-J},$$

$$L = \left(\frac{K}{0.70831}\right)^{-9.642\cos(X_1)} + \log_2\left[\sin(10) - \log_2(10)\right],$$

$$M = \frac{\left(0.59066 - \sqrt{\log(X_1)+10}\right)}{L} + \log(X_1),$$

$$N = \cos(10) - \sqrt{\frac{0.259004}{X_1}} + \log(X_1), \quad O = 1.557408N,$$

$$P = 1.557408\left(O - \sqrt{m}\right), \quad Q = \left(e^{0.671654\tan(X_1)X_1^2}\right),$$

$$R = \left(e^{\tan(X_1)X_1^2 Q}\right), \quad S = R - \frac{\cos\left(e^{\left(\frac{10}{X_1}\right)}\right)}{1.071185}, \quad T = 1.557408\left[\cos(x_1) - S\right]$$

$$U = \frac{e^{\log_{10}(X_1)-T}}{\log_2\left[\sin(10) - \log_2(10)\right]},$$

$$V = U^{-9.642\cos(X_1)},$$

$$W = \log_2\left\{\log_2\left[\log_2\left(\log_{10}(0.98407 - x_1 - 2.142536)\right)\right]\right\} - 3.321928 + V,$$

$$X = \left[0.80812 + \frac{8.036777}{x}\right], \quad Y = 1.557408\left[\cos(x_1) - \sqrt{\log(x_1)+1}\right],$$

$$Z = \left[\cos(x_1) - \frac{e^{\log_{10}(X_1)-Y}}{0.70831}\right]\Big/ W,$$

$$a = \cos(X_1) - \sqrt{\log(X_1) + \frac{0.259004}{X_1}}, \quad b = a\tan\left[e^{\log\left(\frac{10}{X_1}\right)^{0.8901}}\right],$$

$$c = b - \sqrt{\log(X_1) + Z}$$
,

$$d = c \tan\left(e^{\frac{\log_{10}(X_1)}{0.30103}}\right)$$
, $e = F - \left[\log\left(e^{\log_{10}(c)-P}\right)\right]$, $f = e^{[c/(10-X_1)]}$

where, X_1 is the number of particles per unit solid angle (Y), X_2, lab momentum (P_{Lab}), X_3, mass number (A).

Chapter 4

PRESENCE OF MULTIFRACTALITY IN HIGH-ENERGY NUCLEAR COLLISIONS

M. I. Haque[1], M. Tariq[2], Tahir Hussain[3]

[1]Department of Kulliyat, AK Tibbiya College, Aligarh Muslim University, Aligarh, India

[2]Department of Physics, Aligarh Muslim University, Aligarh, India

[3]Department of Applied Physics, Aligarh Muslim University, Aligarh, India

ABSTRACT

In the present study an attempt is made to examine multifractality in multiparticle production in relativistic nuclear collisions; multifractality is investigated in 14.5 A GeV/c^{28}Si-nucleus collisions. For this, G_q-moments are calculated and variations of $\ln\langle G_q \rangle$ with $-\ln\delta\eta$ are looked into. Values of mass exponents, t_q, and generalised dimensions, D_q, are obtained. Analysis of multifractal moments reveals that multiplicity fluctuations are of dynamical nature.

INTRODUCTION

Analysis of high-energy heavy-ion collisions [1] -[4] offers a unique opportunity to investigate occurrence of dynamical fluctuations [5] - [7] in A-A collisions. To understand the real dynamics of multiparticle production, multifractality is envisaged to become an important tool for both theoretician and experimentalist. Intermittency and multifractality in turbulent fluids have been extensively studied [8] . It was suggested that multifractal analysis is carried out by calculating multifractal moments, G_q, of the multiplicity distributions in a given pseudorapidity (η) space. The main purpose of adopting multifractal moments, G_q, approach is to explain multifractality and self-similarity in multiparticle production in relativistic nucleus-nucleus collisions. However,

G_q- moments are greatly influenced by statistical fluctuations in the case of events having lower multiplicities. It is worth mentioning that if the particle production process exhibits self-similar behavior, a modified form of G_q-moment is used by introducing a step function [2] , which leads to power-law dependence on the phase space bin size. Importance of multifractal analysis of high-energy nuclear collision data lies in the fact that these moments can be calculated for the negative values of the order of moments, q, also, whereas factorial moments are defined only for positive integral values of the order of moments. High-energy heavy-ion collisions are considered to be an ideal site for creating the conditions for producing Quark-Gluon Plasma (QGP) and dynamical fluctuations are one of the most reliable signals of QGP formation. Fluctuations in multiplicity and pseudorapidity distributions [9] are the most significant approaches to study nuclear matter produced in these collisions. Using the calculated values of various moments, non-statistical fluctuations in high-energy nuclear interactions can be investigated. It may be noted that multifractality may play an important role for searching the existence of dynamical fluctuations in the multipaticle production.

MATHEMATICAL FORMALISM

In order to study multifractality, a selected pseudorapidity interval, $\Delta\eta$ is partitioned into M bins of equal size $\delta\eta = \Delta\eta/M$. Let n_j be the number of particles lying in j^{th} bin, then multifractal moments, G_q, may be calculated [1] [10] using:

$$G_q = \sum_j (p_j)^q$$

(1)

where quantity p_j is defined as $p_j = n_j/n$ and n be the total number of particles.

In the above expression the summation is carried over non-empty bins only. For a given data sample, averaging is done over all the events comprising the total number of events, N_{evt}, the average value of multifractal moments, $\langle G_q \rangle$ is calculated from:

$$\langle G_q \rangle = \frac{1}{N_{evt}} \sum_1^{N_{evt}} G_q$$

(2)

If rapidity distribution possesses fractal nature, a power-law behavior of $\langle G_q \rangle$ of the following type should be observed over a small pseudorapidity range, $\delta\eta$:

$$\langle G_q \rangle \propto (\delta\eta)^{t_q}$$

$$(3)$$

where t_q are known as mass exponents.

The resulting linear dependence of $\ln\langle G_q \rangle$ on $-\ln\delta\eta$ may be used to determine the values of t_q making use of the following relationship:

$$t_q = \lim_{\delta\eta \to 0} \frac{\Delta\ln\langle G_q \rangle}{\Delta\ln\delta\eta}$$

$$(4)$$

The generalized dimensions, D_q, are considered to contain useful property regarding fractals occurring in multiparticle production in relativistic nuclear collisions. Generalized dimensions are defined as:

$$D_q = \frac{t_q}{q-1}$$

$$(5)$$

Increase in the value of D_q with q is said to describe the pattern as multi fractal, whereas constancy of D_q would point towards monofractality.

EXPERIMENTAL DETAILS

We have analyzed data set comprising 555 events produced in 14.5 A GeV/c ^{28}Si-nucleus collisions. Data sample include collisions with $n_h \geq 0$, where n_h represent the number of charged particles produced with relative velocities, $\beta \leq 0.7$. Experimental results have been compared with the corresponding results for the data generated using Lund model, FRITIOF and Monte Carlo simulation.

RESULTS AND DISCUSSION

Study of $\ln\langle G_q \rangle$ as a Function of $-\ln\delta\eta$

Figure 1 shows the variations of $\ln\langle G_q \rangle$ with $-\ln\delta\eta$ for the experimental data on 14.5 A GeV/c ^{28}Si-nuc- leus collisions; these variations are studied for three groups of targets namely, CNO, emulsion and AgBr groups of nuclei for various classes of interactions. It may be emphasized that multifractal moments show linearly increasing trend with decreasing bin width, $\delta\eta$, for all the three groups of the targets; this linearly increasing behavior which is shown over a large interval of $^{-\ln\delta\eta}$ for positive order of the moments, q, in comparison to the ones for negative q values. For negative values saturating trend is discernible with decreasing $\delta\eta$. The only reason for this behavior appears to be the fact that particle multiplicity will decrease with decreasing bin width, $\delta\eta$ [11] . Multifractal moments for CNO group of targets saturates a little earlier in comparison to those for AgBr nuclei. This linearly increasing nature demonstrated by multifractal moments in the rapidity space indicates the presence of self-similarity. To investigate the dynamical fluctuations using multifractal moments, experimental results are compared with the Monte Carlo generated data sets for 14.5 A GeV/c ^{28}Si-nucleus collisions. Figure 2 compares the behaviors of $^{\ln\langle G_q\rangle}$ vs $-\ln\delta\eta$ plots for the experimental, FRITIOF and Monte Carlo generated data sets. The variation for the MC simulated data is very smooth as compared to those for the experimental and FRITIOF data samples.

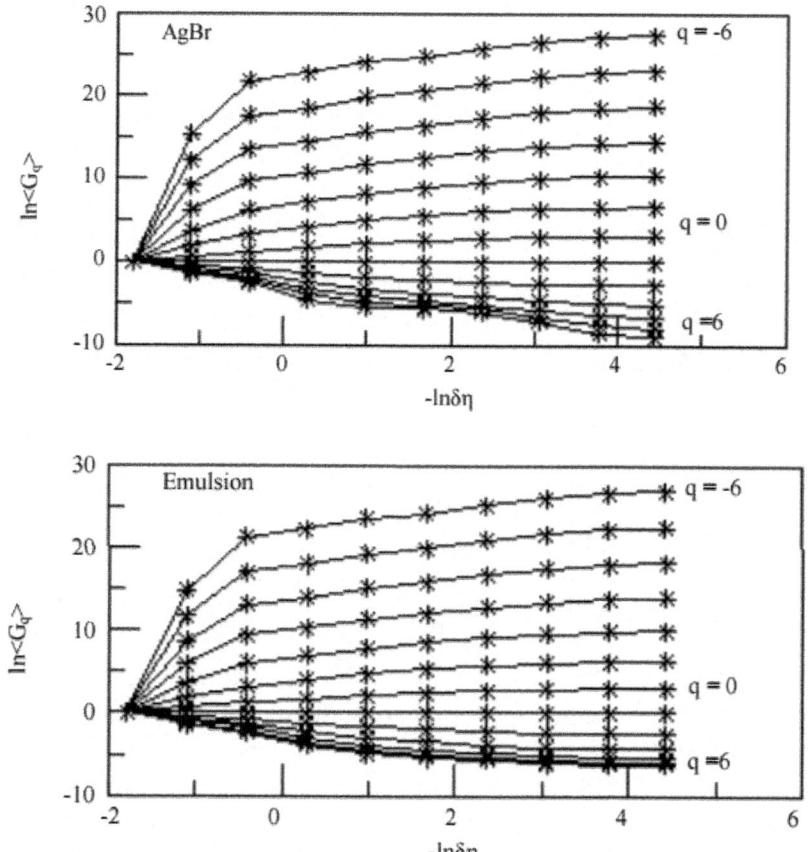

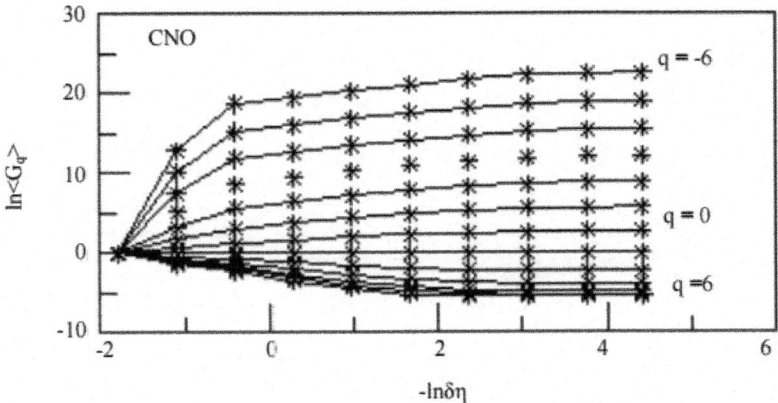

Figure 1: Variations of $\ln\langle G_q \rangle$ with $-\ln\delta\eta$ for the experimental data on 14.5 A GeV/c ^{28}Si-nucleus collisions.

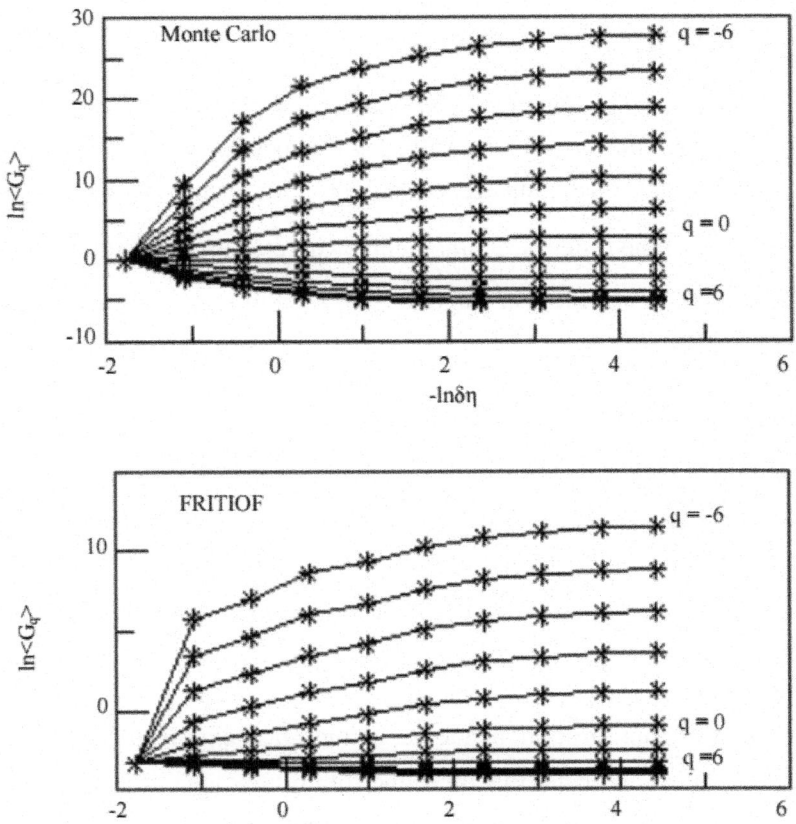

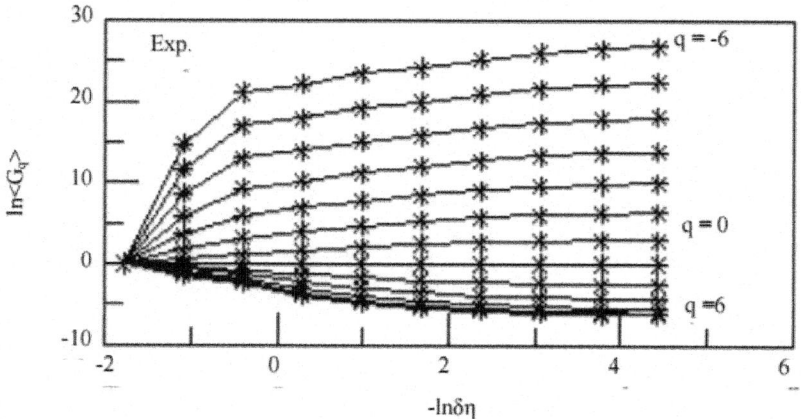

Figure 2: Variations of $\ln\langle G_q \rangle$ with $-\ln\delta\eta$ for the experimental and simulated data samples.

Mass Exponents

Values of the mass exponents, t_q, are determined by fitting the $\ln\langle G_q \rangle$ versus $-\ln\delta\eta$ plots in the region which exhibits linear behavior. The slopes of these fits give the values of the mass exponents. Figure 3 show the variations of t_q, t_q^{stat} and t_q^{dyn} with the order of the moments, where t_q^{stat} represent the slopes for the Monte Carlo generated data and t_q^{dyn} are the dynamical component of the fluctuations; t_q^{dyn}, t_q^{stat}, t_q and q satisfy the following relationship:

$$t_q^{dyn} = t_q - t_q^{stat} + q - 1$$

(5)

Figure 3 shows that the values of t_q increases with increasing q. The increasing trends for positive and negative values of q are quite dissimilar. These values depends on $q > 0$ and $q < 0$. t_q^{dyn} are observed to be quite different from t_q. Figure 4 compares the values of the mass exponents, t_q, for the experimental and FRITIOF generated data. It is noticed that the variations are similar for both the data sets. Again target dependence of the mass exponents, t_q, which are shown in Figure 5, indicates that the values of mass exponents t_q, for $q < 1$, have lower values for the heavier targets as compared to those for the lighter ones, for which $q > 1$.

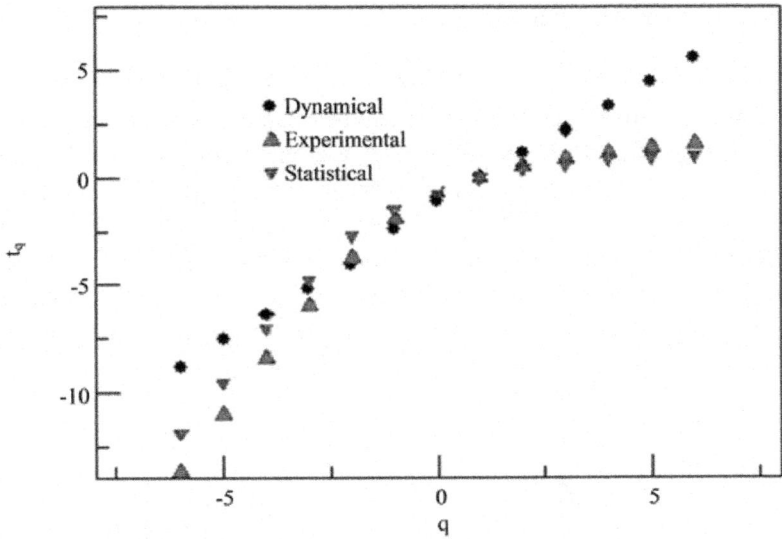

Figure 3: Variations of t_q, t_q^{stat} and t_q^{dyn} with q .

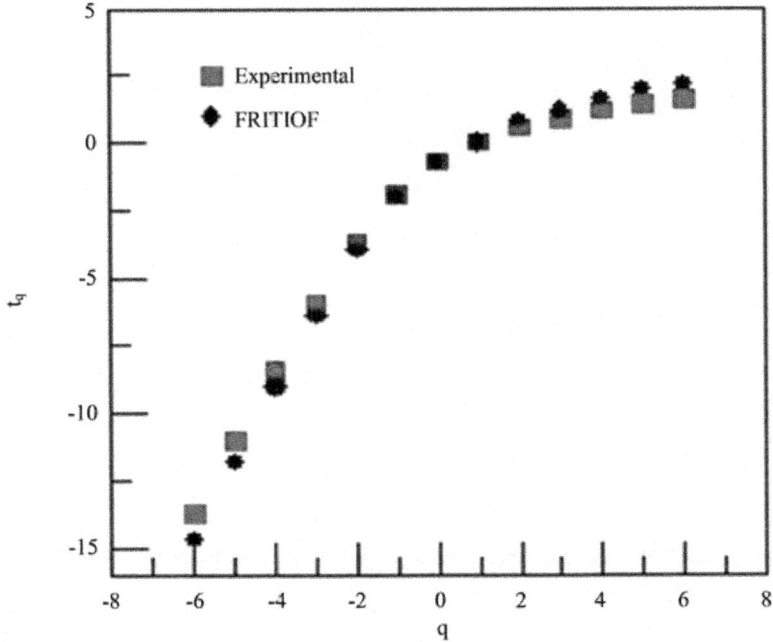

Figure 4: Variations of t_q with q for the experimental and FRITIOF data.

Generalized Dimensions

Figure 6 exhibits the variations of the generalized dimensions, D_q, with q for the three categories of targets. From the figure it is clear that the values of D_q for higher values of q are relatively lower in comparison to those for the lower values of q. For different orders of the moments, q, the values of generalized dimensions is positive and decrease with increasing q. This behavior supports the predictions of the multifractal cascade model. It is clear from the Figure 6 that the generalized dimensions have higher values for heavier targets. The high values of generalized dimensions for heavier targets may be attributed to the fact that multiplicity for heavier targets is higher.

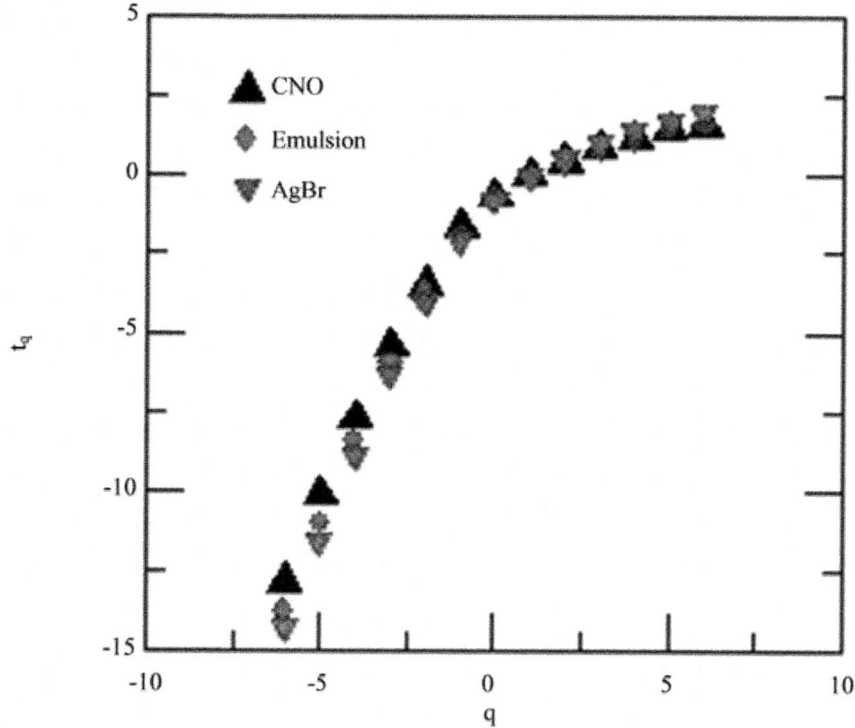

Figure 5: Variations of mass exponents, t_q, with q for the experimental data on 14.5 A GeV/c ^{28}Si-nucleus interactions.

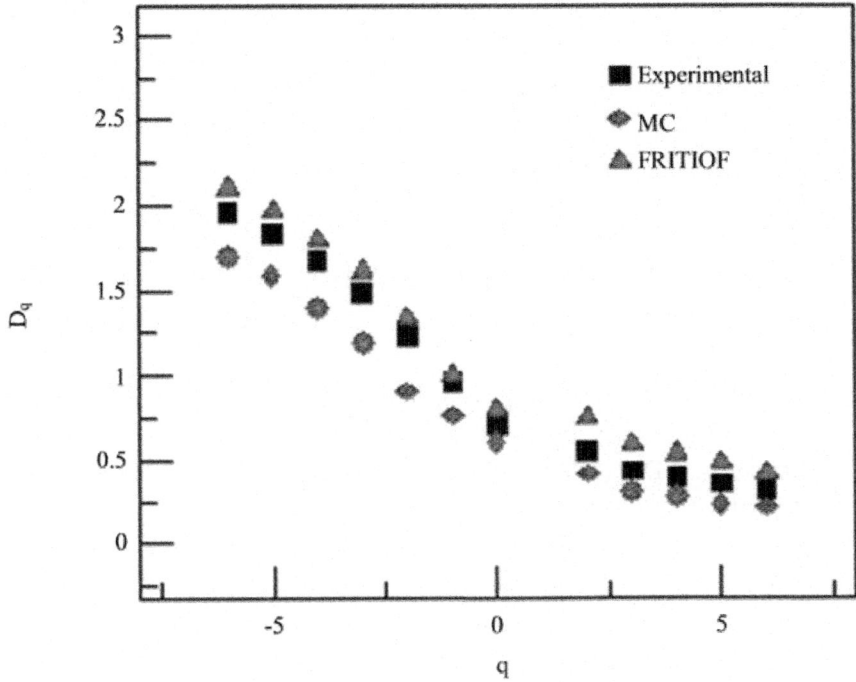

Figure 6: Variations of generalized dimensions, D_q, with q for the experimental, FRITIOF and Monte Carlo generated data.

CONCLUSION

The results of the present study are quite important for drawing meaningful conclusions regarding occurrence of multifractality in multiparticle production in high-energy nucleus-nucleus collisions. It is clearly observed that $\ln\langle G_q \rangle$ first increases linearly with $-\ln\delta\eta$ and then saturates as η resolution increases. The nature of variation clearly hints towards power-law behavior of multifractal moments as a function of $\delta\eta$, thereby indicating the presence of multifractality.

The variations of mass exponents t_q, and generalized dimensions D_q, with q also support the presence of multifractality in the collisions considered in the present study.

REFERENCES

1.   Hwa, R.C. (1990) Physical Review D, 41, 1456.http://dx.doi.org/10.1103/PhysRevD.41.1456

2.   Hwa, R.C. and Pan, J.C. (1992) Physical Review D, 45, 1476.http://dx.doi.org/10.1103/PhysRevD.45.1476

3.   Jain, P.L., Sengupta, K. and Singh, G. (1990) Physics Letters B, 24, 273. Derado, T., et al. (1992) Physics Letters B, 283, 151. Albajar, C., et al. (1992) Zeitschrift für Physik C, 56, 37.

4.   Ghosh, D., et al. (1991) Physics Letters B, 272, 5. Ghosh, D., et al. (1996) Zeitschrift für Physik C, 71, 243.

5.   Schmidt, H.R. and Schukraft, J. (1993) Journal of Physics G: Nuclear and Particle Physics, 19, 1705. http://dx.doi.org/10.1088/0954-3899/19/11/006

6.   Kajantie, K. and Mclerran, L. (1982) Physics Letters B, 119, 203.http://dx.doi.org/10.1016/0370-2693(82)90277-5

7.   Bjorken, J.D. (1983) Physical Review D, 27, 140.http://dx.doi.org/10.1103/PhysRevD.27.140

8.   Mandelbrot, B.B. (1982) The Fractal Geometry of Nature. Freeman, New York.

9.   Muller, B. (1995) Reports on Progress in Physics, 58, 611. Singh, C.P. (1993) Physics Reports, 236, 147.

10.   Chiu, C.B. and Hwa, R.C. (1991) Physical Review D, 43, 100.http://dx.doi.org/10.1103/PhysRevD.43.100

11.   Jain, P.L., Singh, G. and Mukhopadhyay, A. (1992) Physical Review C, 46, 721. http://dx.doi.org/10.1103/PhysRevC.46.721

Chapter 5

DARK MATTER DIRECT-DETECTION EXPERIMENTS

Teresa Marrodán Undagoitia and Ludwig Rauch

Max-Planck-Institut für Kernphysik, Saupfercheckweg 1, D-69117 Heidelberg, Germany

ABSTRACT

In recent decades, several detector technologies have been developed with the quest to directly detect dark matter interactions and to test one of the most important unsolved questions in modern physics. The sensitivity of these experiments has improved with a tremendous speed due to a constant development of the detectors and analysis methods, proving uniquely suited devices to solve the dark matter puzzle, as all other discovery strategies can only indirectly infer its existence. Despite the overwhelming evidence for dark matter from cosmological indications at small and large scales, clear evidence for a particle explaining these observations remains absent. This review summarises the status of direct dark matter searches, focusing on the detector technologies used to directly detect a dark matter particle producing recoil energies in the keV energy scale. The phenomenological signal expectations, main background sources, statistical treatment of data and calibration strategies is discussed.

INTRODUCTION

Overwhelming evidence for gravitational interactions between baryonic and a new form of non-luminous matter can be observed on cosmological as well as astronomical scales. Its nature, however, remains uncertain. It is commonly assumed that elementary particles could be the constituents of this 'dark' matter. Such new particles that could account for dark matter appear in various theories beyond the standard model of particle physics. A variety of experiments have been developed over recent decades, aiming to detect these massive particles via their scattering in a detector medium. Measuring this process would provide information on the dark-matter particle mass and its interaction probability with ordinary matter. The identification of the nature of dark matter would answer one of the most important open questions in physics and would help to better understand the Universe and its evolution. The main goal of this article is to review current and future direct-detection experimental efforts.

This review is organized in the following way. In section 2, the different phenomena indicating the existence of dark matter and possible explanations or candidates emphasizing particle solutions are presented. If, indeed, particles are the answer to the dark matter puzzle, there are three main possibilities for a verification: to produce them at particle accelerators, to look for products of e.g. their self-annihilations at locations with a high dark matter density, or to directly measure their scattering off a detector's target material. This review is dedicated to direct-detection searches for massive particles producing recoil energies in the keV energy scale. The production of dark matter particles at accelerators and searches for indirect signals are discussed only briefly. As the local density and velocity distributions of dark matter are relevant for the interpretation of the experimental results, the main characteristics of the Milky Way halo are presented in section 2. Next, in section 3, the principles of direct detection of WIMPs including the expected signal signatures are explained. Assumptions on particle and nuclear physics aspects which are necessary for the derivation of the results are summarised, and possible interpretations of the results are given. In section 4, a general overview of background sources in direct-detection experiments is given considering different types of radiation and sources of both internal and external contributions to the target material. In section 5, the basic detector technologies are introduced along with their capability to distinguish between signal and background events. Furthermore, statistical methods and the general result of an experiment are discussed. In section 6, the required calibrations to determine the energy scale, energy threshold as well as signal and background regions are detailed. In the main part of this review, section 7, the working principles of different direct detection technologies and the current experimental status are reviewed. Finally, in section 8, the experimental results are summarised, and the prospects for the future are discussed.

THE DARK MATTER PUZZLE

A wealth of observational data from gravitational effects at very different length scales supports the existence of an unknown component in our Universe. After a brief review of these observations ranging from cosmological to Milky Way-sized galaxies, various explanations and elementary-particle candidates are discussed in the following. At the end of the section, possible methods to detect particle dark matter are presented.

Dark matter indications from cosmology and astronomy

Temperature anisotropies in the cosmic microwave background (CMB), precisely measured by WMAP [1] and more recently by the Planck satellite [2], give access to the Universe when it was about 400 000 years old. The power spectrum of temperature fluctuations can be evaluated by a six-parameter model which contains, among others, the baryonic matter, dark matter and dark energy contents of the Universe. This cosmological standard model, which fits the data with high significance, is denoted Λ cold dark matter (ΛCDM) indicating that dark matter with a small random velocity is a fundamental ingredient. The Λ refers to the cosmological constant necessary to explain the current accelerated expansion of the Universe [3]. Oscillations of the baryon–photon fluid in the gravitational potential dominated by CDM density perturbations give rise to the characteristic oscillation pattern in the CMB power spectrum (acoustic peaks). From the relative height of these acoustic peaks, the amount of baryonic matter can be estimated, which allows one to calculate the total dark matter density in the Universe. Present estimates [4] show a flat Universe with $\Omega_{DM} = 0.265$, $\Omega_b = 0.049$ and $\Omega_\Lambda = 0.686$ representing the densities of dark matter, baryonic matter and dark energy, respectively.

In the standard scenario, the anisotropies of the CMB originate from quantum fluctuations during inflation. In order to understand the formation of matter distributions from the time of recombination to the present state, N-body simulations of dark matter particles have been carried out [5]. These simulations [6–8] propagate particles using supercomputers aiming to describe the structure growth, producing a cosmic web ranging from ~10 kpc objects to the largest scales. Meanwhile, these types of simulations reproduce very accurately the measurements made by Galaxy surveys [9–11]. Measurements of the Lyman-α forest [12, 13] and weak lensing [14, 15] confirm the cosmic structure considering not only galaxies and gas clouds but also non-luminous and non-baryonic matter. Large scale simulations, which consider only dark matter, have been used to confirm theories of large scale structure formation which serve as seeds for Galaxy and cluster formation. Recently, gas and stars have been included in the simulations and it is shown that they can significantly alter the distribution of the dark matter component on small scales [16].

A further hint for the existence of dark matter arises from gravitational lensing measurements [15]. This effect discussed by Albert Einstein [17] in 1936 and later by Fritz Zwicky [18] occurs when a massive object is in the line of sight between the observer at the Earth and the object under study.

The light-rays are deflected through their path due to the gravitational field resulting, for example, in multiple images or a deformation of the observable's image (strong and weak lensing, respectively). The degree of deformation can be used to reconstruct the gravitational potential of the object that deflects the light along the line of sight. From various observations it has been found that the reconstructed mass using this method is greater than the luminous matter, resulting in very large mass to light ratios (from a few to hundreds). Gravitational lensing has also been applied in Galaxy-cluster collisions to reconstruct the mass distributions in such events where mass to light ratios of >200 are measured. In some examples [19–21] and in an extensive study of 72 cluster collisions [22], the reconstructed gravitational centers appear clearly separated from the main constituent of the ordinary matter, i.e. the gas clouds which collide and produce detectable x-rays. This can be interpreted as being due to dark matter haloes that continue their trajectories independently of the collision. An upper limit to the self-interaction cross-section for dark matter can be derived from these observations [23].

Indications for non-luminous matter appear in our Universe also at smaller scales. Historically, the first indications for dark matter arose from astronomical observations. In order to explain measurements of the dynamics of stars in our Galaxy, the word 'dark matter' was already used by Kapteyn [24] in 1922 but it was not the correct physical explanation of the observed phenomenon. The first evidence of dark matter in the present understanding was the measurement of unexpectedly high velocities of nebulae in the Coma cluster which brought Zwicky [25] to the idea that a large amount of dark matter could be the explanation for the unexpected high velocities. In 1978, Rubin et al [26] found that rotation velocities of stars in galaxies stay approximately constant with increasing distance to their galactic center. This observation was in contradiction with expectation, as objects outside the visible mass distribution should have velocities $v \propto 1/\sqrt{r}$ following Newtonian dynamics. A uniformly distributed halo of dark matter could explain both the velocities in clusters and the rotation velocities of objects far from the luminous matter in galaxies (e.g. [27]).

The nature of dark matter: possible explanations and candidates

A plausible solution to describe some of the astronomical measurements mentioned in section2.1 is a modification of gravitation laws to accommodate the observations. Such modified Newtonian dynamic models like MOND [28] or its relativistic extension TeVeS [29] can, for instance, successfully describe rotational velocities measured in galaxies. However, MOND fails or needs unrealistic parameters to fit observations on larger scales such as structure formation or the CMB structure and violates fundamental laws such as momentum conservation and the cosmological principle [30]. While TeVeS can solve some of the conceptual problems of MOND, the required parameters seem to generate an unstable Universe [31] or fail to simultaneously fit lensing and rotation curves [32].

Massive astrophysical compact halo objects (MACHOs) have also been considered as a possible explanation for the large mass to light ratios detected in the astronomical observations described in the previous section. These objects could be neutron stars, black holes, brown dwarfs or unassociated planets that would emit very little to no radiation. Searches for such objects using gravitational microlensing [33] towards the Large Magellanic Cloud have been performed [34]. Extrapolations to the Galactic dark matter halo showed that MACHOs can make up about 20% of the dark matter in our Galaxy and that a model with MACHOs accounting entirely for the dark-matter halo is ruled out at 95% confidence level [34]. The baryonic nature of dark matter is actually also ruled out by big-bang nucleosynthesis (BBN). The abundance of light elements predicted by BBN depends on the baryon density and, in fact, measurements constrain the baryon density to a value around $\Omega_b = 0.04$ [35] close to the value derived from CMB.

A more common ansatz is to assume that dark matter is made out of massive neutral particles featuring a weak self-interaction. From the known particles in the standard model, only the neutrino could be considered. Due to its relativistic velocity in the early Universe, the neutrino would constitute a hot dark matter candidate. Cosmological simulations have shown, however, that a Universe dominated by neutrinos would not be in agreement with the observed clustering scale of galaxies [36]. Furthermore, due to the fermionic character of neutrinos, their occupation number is constrained by the Fermi–Boltzmann distribution, so they cannot account for the observed dark-matter density in halos [37]. Sterile neutrinos are hypothetical particles which were originally introduced to explain the smallness of the neutrino masses [38]. Additionally, they provide a viable dark matter candidate. Depending on their production mechanism, they would constitute a cold (non-relativistic at all times) or a warm (relativistic only in an early epoch) dark matter candidate [39, 40]. Possible masses, which are not yet constrained by x-ray measurements or the analysis of dwarf spheroidal galaxies, range from 1 keV to tens of keV. Given this very low mass, and the low interaction strength, the existence of sterile neutrinos is not tested by direct detection experiments. An indication could, for example, arise from the x-ray measurement of the sterile neutrino decay via the radiative channel $N \to \nu\gamma$ [41].

Models beyond the standard model of particle physics suggest the existence of new particles which could account for the dark matter. If such hypothetical particles were stable, neutral and had a mass from below GeV/c^2 to several TeV/c^2, they could be the weakly interacting massive particles (WIMPs). The standard production mechanism for WIMPs assumes that in the early Universe these particles were in equilibrium with the thermal plasma [42]. As the Universe expanded, the temperature of the plasma became lower than the WIMP mass resulting in the decoupling from the plasma. At this freeze-out temperature, when the WIMP annihilation rate was smaller than the Hubble expansion rate, the dark matter relic density was reached. The cross-section necessary to observe the current dark matter density is of the order of the weak interaction scale. It appears as a great coincidence that a particle interacting via the weak force would produce the right relic abundance and, therefore, the WIMP is a theoretically well motivated dark matter candidate.

Supersymmetry models [43] are proposed as extensions of the standard model of particle physics to solve the hierarchy problem as well as the unification of weak, strong and electromagnetic interactions. In this model, a whole new set of particles are postulated such that for each particle in the standard model there is a supersymmetric partner. Each particle differs from its partner by 1/2 in spin and, consequently, bosons are related to fermions and vice versa. The neutralino, the lightest neutral particle which appears as a superposition of the partners of the standard model bosons, constitutes an example of a new particle fulfilling the properties of a WIMP. The typical masses predicted for the neutralino range from few GeV/c^2 to several TeV/c^2. A WIMP candidate appears also in models with extra–dimensions. In such models N spatial dimensions are added to the $(3 + 1)$ space–time classical ones. They appeared already around 1920 to unify electromagnetism with gravity. The lightest stable particle is called 'lightest Kaluza particle' and constitutes also a good WIMP candidate [44, 45].

Among the non-WIMP candidates, 'superheavy dark matter' or 'WIMPzillas' are postulated to explain the origin of ultra-high-energy cosmic rays [46]. At energies close to 10^{20} eV, cosmic protons can interact with the CMB and, thus, their mean free path is reduced resulting in a suppressed measured flux [47, 48]. Experimental results include, however, the detection of a few events above the expected cut-off, motivating a superheavy dark matter candidate. Decays of these non-thermally produced [49] superheavy particles with masses of $(10^{12} - 10^{16})$ GeV/c^2 could account for the observations, being at the same time responsible for the dark matter in the Universe.

Finally, a very well-motivated particle and dark matter candidate is the axion. In the standard model of particle physics, there is no fundamental reason why QCD should conserve P and CP. However, from the experimental bound on the neutron electric dipole moment [50], very small values of P and CP violation can be derived. In order to solve this so-called 'strong CP-problem' [51], a new symmetry was postulated [52] in 1977. When this symmetry is spontaneously broken, a massive particle, the axion, appears. The axion mass and the coupling strength to ordinary matter are inversely proportional to the breaking scale which was originally associated to the electroweak scale. This original axion model is ruled out by laboratory experiments [53]. Cosmological and astrophysical results also provide very strong bounds on the axion hypothesis [51]. There exist, however, further 'invisible' axion models in which the breaking scale is a free parameter, KSVZ [54, 55] and DFSZ [56, 57], and still provide a solution to the CP-problem. Invisible axions or axion-like particles, would have been produced non-thermally in the early Universe by mechanisms like the vacuum realignment [58, 59] for example, giving the right dark matter abundance. The resulting free streaming length would be small and, therefore, these axions are a 'cold' candidate. For certain parameters, axions could account for the complete missing matter [60].

Sterile neutrinos, WIMPs, superheavy particles and axions are not the only particle candidates proposed. The candidates mentioned above arise from models that were proposed originally with a different motivation and not to explain dark matter. The fact that the models are motivated by different unresolved observations strengthens the relevance of the predicted dark matter candidate. A more comprehensive review on dark matter candidates can be found for example in [61]. This article focuses on the direct detection of WIMPs and just some brief information on searches for particles that would induce an electronic recoil (e.g. axion-like particles) will be given in the following.

Searches for Dark Matter Particles

The particle dark matter hypothesis can be tested via three processes: the production at particle accelerators, indirectly by searching for signals from annihilation products, or directly via scattering on target nuclei. Figure 1 shows a schematic representation of the possible dark matter couplings to a particle, P, of ordinary matter.

While the annihilation of dark matter particles (downwards direction) could give pairs of standard model particles, the collision of electrons or protons at colliders could produce pairs of dark matter particles. In this section the production and indirect detection methods as well as the current status of searches are briefly summarised. The subsequent sections and main part of this review are then devoted to the direct detection of dark matter, $\chi P \to \chi P$ (horizontal direction in figure 1).

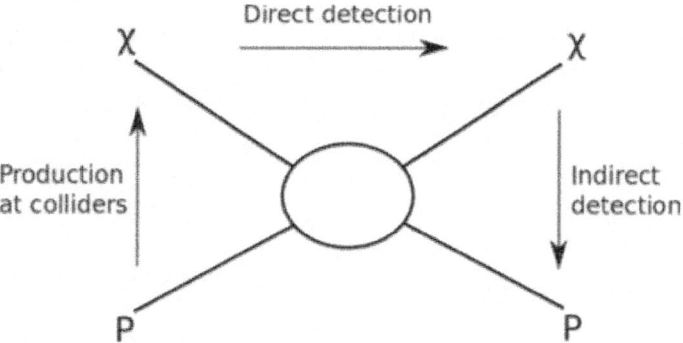

Figure 1: Schematic showing the possible dark matter detection channels.

Since the start of the Large Hadron Collider (LHC) at CERN in 2008, the CMS [62] and ATLAS [63] experiments have searched for new particles in proton–proton collisions at a center-of-mass energy of 7 TeV. Besides the discovery of the Higgs particle [64, 65], CMS and ATLAS have studied a number of new particle signatures by scanning the parameter space of different supersymmetric and extra-dimensions models. The presence of a dark matter particle would only be inferred by observing events with missing transferred momentum and energy. Therefore, events with, e.g., an energetic jet and an imbalanced momentum transfer are selected for analysis. Reactions of the type

$$pp \rightarrow \chi\bar{\chi} + x \tag{1}$$

are probed, x being a hadronic jet, a photon or a leptonically decaying Z or W boson. The results obtained so far are consistent with the standard model expectations (see for example [66–68]) but further searches will be performed in the next few years for higher center-of-mass energy. The derived bounds can be translated into limits on the cross-section for a given particle mass. Bounds arising from accelerator searches are most constraining below ~4 GeV and ~700 GeV for spin-independent (SI) and spin-dependent (SD) (proton coupling, see section 3.2) interactions, respectively [66]. However, a direct comparison of these experimental results to other detection methods is, in general, model dependent (see the discussion in section 8).

Dark matter particles can gravitationally accumulate in astrophysical objects such as stars, galaxies or our Sun. The most favoured sources to search for indirect signals are the galactic centre and halo, close Galaxy clusters or dwarf galaxies also called dwarf spheroidals. The latter are very popular locations due to their large measured mass to light ratio and their small background. Due to the increased dark-matter density, an enhanced self-annihilation, scattering or decay into standard model particles could produce a measurable particle flux (see [69] for a detailed discussion). The measurement of these secondary particles is a further detection mechanism usually denoted as 'indirect detection'. Examples of possible annihilation channels are

$$\chi \bar{\chi} \rightarrow \gamma\gamma, \ \gamma Z, \ \gamma H \quad \text{or} \tag{2}$$

$$\chi \bar{\chi} \rightarrow q\bar{q}, \ W^- W^+, \ ZZ. \tag{3}$$

Some of the products decay further into $e^- e^+$, $p\bar{p}$·γ-rays and neutrinos. A second mechanism to generate charged (anti-) particles, photons or neutrinos from dark matter is given by its decay. In contrast to self-annihilation processes, where the production rate shows a quadratic dependence of the dark matter density, decaying dark matter scales only linearly (e.g. [70]). In addition, dark matter particles might be gravitationally captured inside the Sun due to the elastic scattering with its nuclei. The annihilation of captured dark-matter particles can produce neutrinos which can propagate out of the Sun and might be detectable with Earth-based neutrino telescopes. Note that the total number of captured particles is less affected by uncertainties of the dark matter halo since this process lasts for billions of years and dark matter density variations are averaged out [70].

Produced charged particles are deflected in the interstellar magnetic fields, losing the information on their origin. Due to their charge neutrality, γ-rays and neutrinos point, instead, to the source where they were produced. While neutrinos travel unaffected from the production source, γ-rays can be affected by absorption in the interstellar medium.

Imaging atmospheric Cherenkov telescopes for TeV γ-ray detection can look specifically in the direction of objects where a large amount of dark matter is expected. Either a γ-flux in dwarf galaxies or Galaxy clusters, or mono-energetic line signatures are searched for. So far no significant signal from dark matter annihilations has been observed, and upper limits are derived by the MAGIC [71, 72], HESS [73, 74] and VERITAS [75, 76] telescopes. Indirect searches can be also performed by satellite-based instruments capable of detecting low-energy γ-rays (approx. 20 MeV–300 GeV) like Fermi-LAT [77]. Although some gamma-ray features identified in the Fermi data are intriguing (for example [78–80]), in the publication of early 2015 by the Fermi collaboration [81] no evidence for a dark-matter signal is found. One of the strongest and most robust constraints can be derived by the Fermi-LAT observation of dwarf spheroidal satellite galaxies of the Milky Way as those are some of the most dark-matter-dominated objects known [82]. Consequently, conservative limits on the annihilation cross-section of dark matter particles ranging from a few GeV to a few tens of TeV are derived. In the energy region of (0.1–10) keV, x-ray satellites such as XMM-Newton and Chandra provide data to search for indirect dark matter signals. In 2014, an unexpected line at 3.5 keV was found in the data recorded by both satellites [83, 84]. This signal can be interpreted by a decay of dark matter candidates, for instance, from sterile neutrinos or axions [85–88]. Other astrophysical explanations have been, however, proposed and thus, the origin of the signal remains controversial (see e.g. [89–91]). Large neutrino detectors like Ice Cube or Super-Kamiokande are able to search for dark matter annihilations into neutrinos. No evidence for such a signal has been observed, resulting in constraints on the cross-section [92, 93]. Finally, also charged particles like protons, antiprotons, electrons and positrons can be detected by satellites. Measurements on the steadily increasing positron fraction from 10 to ~250 GeV by Pamela [94] and AMS [95] raise discussions on its possible dark matter origin. However, given that such a spectrum could be also described by astrophysical objects like pulsars (rapidly rotating neutron stars) or by the secondary production of e^+ by the collision of cosmic rays with interstellar matter [96], this cannot be considered as a clear indication of a dark matter signal.

PRINCIPLES OF WIMP DIRECT DETECTION

Large efforts have been pursued to develop experiments which are able to directly test the particle nature of dark matter. The aim is to identify nuclear recoils produced by the collisions between the new particles and a detector's target nuclei. The elastic scattering of WIMPs with masses of (10–1000) GeV/c^2 would produce nuclear recoils in the range of (1–100) keV [97]. To unambiguously identify such low-energy interactions, a detailed knowledge on the signal signatures, the particle physics aspects and nuclear physics modelling is mandatory. Furthermore, for the calculation of event rates in direct-detection experiments, the dark matter density and the halo velocity distribution in the Milky Way are required. This section is devoted to reviewing all these aspects, focusing on WIMP dark matter, whereas non-WIMP candidates are briefly discussed in section 3.3.

Experimental signatures of dark matter

The signature of dark matter in a direct-detection experiment consists of a recoil spectrum of single-scattering events. Given the low interaction strength expected for the dark matter particle, the probability of multiple collisions within a detector is negligible. In the case of a WIMP, a nuclear recoil is expected [98]. The differential recoil spectrum resulting from dark matter interactions can be written, following [97], as:

$$\frac{dR}{dE}(E, t) = \frac{\rho_0}{m_\chi \cdot m_A} \cdot \int v \cdot f(\mathbf{v}, t) \cdot \frac{d\sigma}{dE}(E, v) \, d^3v,$$

(4)

where m_χ is the dark matter mass and $\frac{d\sigma}{dE}(E, v)$ its differential cross-section. The WIMP cross-section σ and m_χ are the two observables of a dark matter experiment. The dark matter velocity v is defined in the rest frame of the detector and m_A is the nucleus mass. Equation (4) shows explicitly the astrophysical parameters, the local dark matter density ρ_0 and $f(\mathbf{v}, t)$, which accounts for the WIMP velocity distribution in the detector reference frame. This velocity distribution is time dependent due to the revolution of the Earth around the Sun. Based on equation (4), detection strategies can exploit the energy, time or direction dependences of the signal.

The most common approach in direct-detection experiments is the attempt to measure the energy dependence of dark matter interactions. According to [97], equation (4) can be approximated by

$$\frac{dR}{dE}(E) \approx \left(\frac{dR}{dE}\right)_0 F^2(E) \exp\left(-\frac{E}{E_c}\right),$$

(5)

where $\left(\frac{dR}{dE}\right)_0$ denotes the event rate at zero momentum transfer and E_c is a constant parameterizing a characteristic energy scale which depends on the dark matter mass and target nucleus [97]. Hence, the signal is dominated at low recoil energies by the exponential function. $F^2(E)$ is the form-factor correction which will be described in more detail in section 3.2.

Another possible dark matter signature is the so-called 'annual modulation'. As a consequence of the Earth rotation around the Sun, the speed of the dark matter particles in the Milky Way halo relative to the Earth is largest around 2 June and smallest in December. Consequently, the amount of particles able to produce nuclear recoils above the detectors' energy threshold is also largest in June [99]. As the amplitude of the variation is expected to be small, the temporal variation of the differential event rate can be written, following [100], as

$$\frac{dR}{dE}(E, t) \approx S_0(E) + S_m(E) \cdot \cos\left(\frac{2\pi(t - t_0)}{T}\right),$$

(6)

where t_0 is the phase which is expected at about 150 d and T is the expected period of one year. The time-averaged event rate is denoted by S_0, whereas the modulation amplitude is given by S_m. A rate modulation would, in principle, enhance the ability to discriminate against background and help to confirm a dark matter detection.

Directionality is another dark matter signature which can be employed for detection as the direction of the nuclear recoils resulting from WIMP interactions has a strong angular dependence [101]. This dependence can be seen in the differential rate equation when it is explicitly written as a function of the angle γ, defined by the direction of the nuclear recoil relative to the mean direction of the solar motion

$$\frac{dR}{dE\, dcos\gamma} \propto \exp\left[\frac{-\left[(v_E + v_\odot)\cos\gamma - v_{min}\right]^2}{v_c^2}\right].$$

(7)

In equation (7), v_E represents the Earth's motion, $v_\odot$ the velocity of the Sun around the Galactic centre, v_{min} the minimum WIMP velocity that can produce a nuclear recoil of an energy E and v_c the halo circular velocity $v_c = \sqrt{3/2}\, v_\odot$. The integrated rate of events scattering in the forward direction will, therefore, exceed the rate for backwards scattering events by an order of magnitude [101]. An oscillation of the mean direction of recoils over a sidereal day is also expected due to the rotation of the Earth and if the detector is placed at an appropriate latitude. This directional signature allows one to discriminate potential backgrounds [102]. A detector able to determine the direction of the WIMP-induced nuclear recoil would provide a powerful tool to confirm the measurement of dark matter particles. Such directional searches are summarized in section 7.6.

Cross-sections and nuclear physics aspects

To interpret the data of dark matter experiments, further assumptions on the specific particle-physics model as well as on the involved nuclear-physics processes have to be made. This section summarises the most common interactions between dark matter particles and the target nucleons.

For WIMP interactions that are independent of spin, it is assumed that neutrons and protons contribute equally to the scattering process (isospin conservation). For sufficiently low momentum transfer q, the scattering amplitudes of each nucleon add in phase and result in a coherent process. For SD interactions, only unpaired nucleons contribute to the scattering. Therefore, only nuclei with an odd number of protons or neutrons are sensitive to these interactions. In this case, the cross-section is related to the quark spin content of the nucleon with components from both proton and neutron couplings.

When the momentum transfer is such that the particle wavelength is no longer large compared to the nuclear radius, the cross-section decreases with increasing q. The form factor F accounts for this effect and the cross-section can be expressed as: $\sigma \propto \sigma_0 \cdot F^2$, where σ_0 is the cross-section at zero momentum transfer. In general, the differential WIMP–nucleus cross section, $d\sigma/dE$ shown in equation (4), can be written as the sum of an SI contribution and an SD one

$$\frac{d\sigma}{dE} = \frac{m_A}{2\mu_A^2 v^2} \cdot \left(\sigma_0^{SI} \cdot F_{SI}^2(E) + \sigma_0^{SD} \cdot F_{SD}^2(E) \right).$$

(8)

The WIMP–nucleus reduced mass is described by μ_A. For SI interactions, the cross-section at zero momentum transfer can be expressed as

$$\sigma_0^{SI} = \sigma_p \cdot \frac{\mu_A^2}{\mu_p^2} \cdot \left[Z \cdot f^p + (A - Z) \cdot f^n \right]^2$$

(9)

where $f^{p,n}$ are the contributions of protons and neutrons to the total coupling strength, respectively, and μ_p is the WIMP–nucleon reduced mass. Usually, $f^p = f^n$ is assumed and the dependence of the cross-section with the number of nucleons A takes an A^2 form. The form factor for SI interactions is calculated assuming the distribution of scattering centres to be the same as the charge distribution derived from electron scattering experiments [97]. Commonly, the Helm parameterization [103] is used to describe the form factor. Recent shell-model calculations [104] show that the derived structure factors are in good agreement with the classical parameterization.

To visualize the effect of the target isotope and the form-factor correction, figure 2 (left) shows the event rate given in number of events per keV, day and kg (equation (4)) for SI interactions in different target materials: tungsten in green, xenon in black, iodine in magenta, germanium in red, argon in blue and sodium in grey. A WIMP mass of 100 GeV/c^2 and a cross-section of 10^{-45} cm^2 are assumed for the calculation. In these curves both the A^2 dependence of the cross-section and the form-factor correction affect the shape of the energy spectrum. Heavier elements profit from the A^2 enhancement with a higher event rate at low deposited energies but the coherence loss due to the form factor suppresses the event rate especially at higher recoil energies. Therefore, for lighter targets a low energy threshold is of less relevance than for the heavier ones. Figure 2 (right) shows separately the WIMP mass and the form-factor effect on the differential event rate without considering the nuclear recoil acceptance and the energy threshold of the detector. Solid lines show the expected rates for a 100 GeV/c^2 WIMP as in the left figure for a heavy and a light target as indicated in green (tungsten) and blue (argon), respectively. In comparison to the heavy WIMP mass the rates for a 25 GeV/c^2 dark matter particle (dashed line) drop steeper as the momentum transfer is smaller. The form factor correction for a heavy target is more important than for light targets. This can be seen by the dotted lines representing rates for a 100 GeV/c² WIMP, calculated without the form-factor correction.

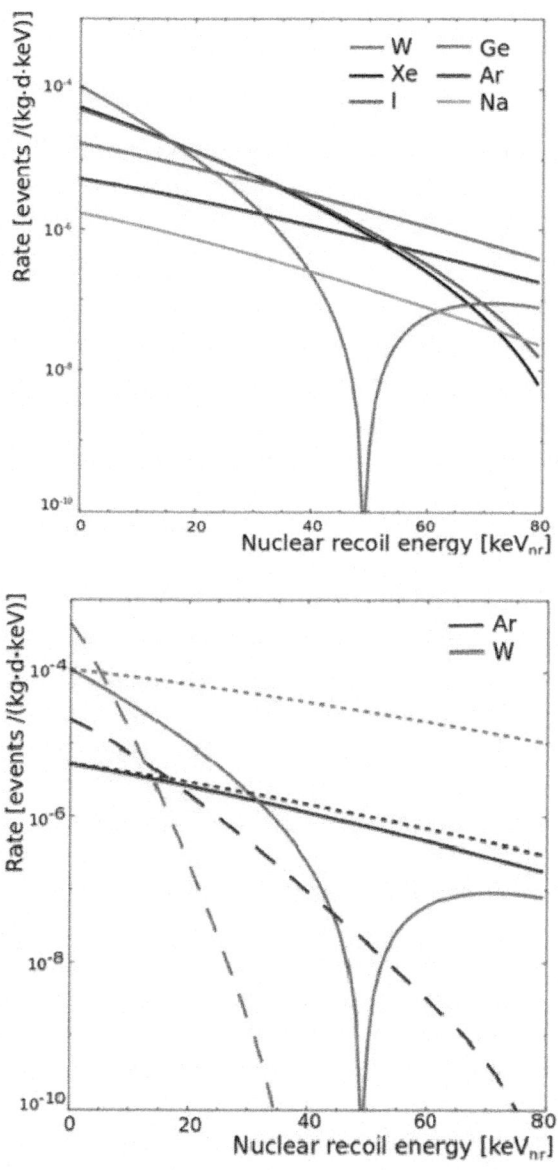

Figure 2: (Left) Differential event rate for the direct detection of a 100 GeV/c² WIMP with a cross-section of 10^{-45} cm² in experiments using tungsten (green), xenon (black), iodine (magenta), germanium (red), argon (blue) and sodium (grey) as target materials. (Right) The event rate is shown for a heavy and a light target as indicated in

green (tungsten) and blue (argon), respectively, showing the effect of neglecting the form factor correction (dotted line) and the effect of a lower WIMP mass of 25 GeVc^2 (dashed line).

For SD interactions, the form factor is written in terms of the spin structure function whose terms are determined from nuclear shell model calculations [105, 106]. A common practice is to express the cross-section for the interaction with protons and with neutrons

$$\sigma_0^{SD} = \frac{32}{\pi} \mu_A^2 \cdot G_F^2 \cdot \left[a_p \cdot \langle S^p \rangle + a_n \cdot \langle S^n \rangle \right]^2 \cdot \frac{J+1}{J},$$

(10)

where G_F^2 is the Fermi coupling constant, J the total nuclear spin and $a_{p,n}$ the effective proton (neutron) couplings. The expectation value of the nuclear spin content due to the proton (neutron) group is denoted by $\langle S^{p,n} \rangle$. New calculations performed in [107] use chiral effective-field theory (EFT) currents to determine the couplings of WIMPs to nucleons up to the leading two-nucleon currents. This method yields an improved agreement between the calculated and measured energy spectra of the considered nuclei as well as the ordering of the nuclear levels (e.g. [108]). These calculations have been used to calculate the couplings for the most relevant isotopes in direct detection experiments: ^{129,131}Xe, ^{127}I, ^{73}Ge, ^{19}F, ^{23}Na, ^{27}Al and ^{29}Si.

In the context of a non-relativistic EFT for WIMP-like interactions, a more detailed formulation of possible couplings from dark matter to baryons has been proposed [109–111] and is applied by some experiments [112]. Instead of the classical two (SI and SD) couplings, six possible nuclear response-functions are assumed which are described by 14 different operators. In this model, the nucleus is not treated as a point-like particle; instead, its composite nature is reflected. Thus, the spin response function is split in transverse and longitudinal components and new response functions arise from the intrinsic velocities of the nucleons. Note that the form factor F, as introduced above, tries to account for the finite spatial extend of the nuclear charge and spin densities. This correction, however, is only approximate. The EFT operators are constructed by four three-vectors $i\frac{\vec{q}}{m_N}$, $\vec{v}^\perp$, $\vec{S}_N$, $\vec{S}_\chi$ which describe the momentum transfer q scaled with the nucleon mass m_N, the WIMP–nucleon relative velocity $\vec{v}^\perp$, the spin of the nucleus $\vec{S}_N$ and the possible spin of the dark matter particle $\vec{S}_\chi$, respectively. The standard SI (equation (9)) and SD (equation (10)) interactions are described by operators $\mathcal{O}_1$ and $\mathcal{O}_4$ with 1 being the identity matrix

$$\mathcal{O}_1 = 1_\chi 1_N, \quad \mathcal{O}_4 = \vec{S}_\chi \cdot \vec{S}_N.$$

(11)

The SI interactions are, furthermore, decomposed into two longitudinal components and a transversal spin component, as in general interactions do not couple to all spin projections symmetrically. New operators arise also by a direct velocity dependence. The impact of the detailed EFT approach on the dark matter limits in comparison to the conventional SI/SD interaction has been calculated in [112] and shows that, in some cases, the compatibility of results among experiments using different targets is significantly affected. Furthermore, destructive interference effects among operators can weaken standard direct-detection exclusion limits by up to one order of magnitude in the coupling constants [113]. This approach not only generalizes the traditional SI and SD parameter space but also allows one to constrain, in an easier way, dark matter models due to the variety of constrained operators.

Other interpretations

The previous section describes a model where dark matter particles scatter off the target nucleus producing nuclear recoils; however, various other models exist. This section briefly summarises a selection of alternative dark matter interactions for which experiments have derived results.

An extension of the standard elastic scattering off nuclei is an inelastic scattering off the WIMP, which was motivated to solve discrepancies among experimental results [114]. In this approach, WIMPs are assumed to only scatter off nuclei by simultaneously getting excited to a higher state with an energy δ above the ground state. The elastic scattering would be, in this case, highly suppressed or even forbidden. The energy spectrum is suppressed at low energies due to the velocity threshold for the inelastic scattering process. Experimental constraints on this model have been shown e.g. in [115–117]. Another possibility is the inelastic WIMP–nucleus scattering in which the target nucleus is left in a low-lying nuclear excited state [118]. The signal would have a signature of a nuclear recoil followed by a γ-ray from the prompt de-excitation of the nucleus. As an example, the inelastic structure functions have been calculated for xenon in [119] and are used in [120] to derive the corresponding exclusion limits for this process.

In contrast to interactions with nucleons, various models allow a dark matter scattering off electrons. For instance, sub-GeV dark matter particles could produce detectable ionization signals [121] and, indeed, limits have been derived for such candidates [122]. Furthermore, if new forms of couplings are introduced to mediate the dark matter–electron interactions, further models become viable. By assuming an axial-vector coupling [123], the dark matter–lepton interactions dominate at tree level and cannot be probed by dark matter–baryon scattering. Furthermore, models such as kinematic-mixed mirror dark matter [124] or luminous dark matter [125] also predict interactions with atomic electrons.

New couplings are also introduced to mediate interactions of axion-like particles (ALPs) with electrons via the axioelectric (also photoelectric-like) or Primakov processes [126] (see section2.2). These processes, invoked by sufficiently massive particles in direct detection experiments, involve only the emission of electrons and x-rays and therefore cannot be separated from the experimental electronic recoil background. Nevertheless, bounds on these models have been derived from data of various experiments [127–130]. The same interactions are assumed for bosonic super-weakly interacting massive dark matter candidates [131] but their electronic recoil energy scale is in general higher and limits are derived in [132].

Distribution of dark matter in the Milky Way

The dark matter density in the Milky way at the position of the Earth and its velocity distribution are astrophysical input parameters, needed to interpret the results of direct-detection experiments. In this section, the parameters of the standard halo model typically used to derive the properties of dark matter interactions, their uncertainties and the differences in modelling the dark-matter halo itself are summarised.

It is common to assume a local dark matter density of 0.3 GeV cm^{-3} which results from mass modelling of the Milky Way, using parameters in agreement with observational data [133]. However, depending on the profile model used for the halo, a density range from (0.2–0.6) GeV cm^{-3} can be derived (see [134] for a review on this topic).

The dark matter velocity profile is commonly described by an isotropic Maxwell–Boltzmann distribution

$$f(\mathbf{v}) = \frac{1}{\sqrt{2\pi}\sigma} \cdot \exp\left(-\frac{|\mathbf{v}|^2}{2\sigma^2} \right)$$

(12)

which is truncated at velocities exceeding the escape velocity. Here, the dispersion velocity σ is related to the circular velocity via $\sigma = \sqrt{3/2}\, v_c$. A standard value of $v_c = 220$ km s^{-1} is used for the local circular speed. This value results from an average of values found in different analyses [135]. More recent studies using additional data and/or different methods, find velocities ranging from (200 ± 20) km s^{-1} to (279 ± 33) km s^{-1} [133]. Finally, the escape velocity defines a cut-off in the description of the standard halo profile. The commonly used value of 544 km s^{-1}is the likelihood median calculated using data from the RAVE survey [136]. The 90% confidence interval contains velocities from 498 km s^{-1} to 608 km s^{-1}. These large ranges of possible values for the dark matter density, circular speed and escape velocity illustrate that the uncertainties in the halo modelling are significant. The GAIA satellite1 , in orbit since January 2014, has been designed to measure about a billion stars in our Galaxy and throughout the Local Group. These unprecedented positional and radial velocity measurements will reduce the uncertainties on the local halo model of the Milky Way.

Not only do the parameters of the dark matter halo show uncertainties but also modelling the halo itself inherits strong assumptions. A sharp truncation of the assumed Maxwell–Boltzmann distribution at the escape velocity has to be unphysical, which motivated the idea of King models (e.g. [137, 138]) trying to account naturally for the finite size of the dark matter halo. It is also possible that the velocity distribution is anisotropic, giving rise to triaxial models, allowing different velocities in each dimension of the velocity vector (e.g. [139, 140]). If the dark matter halo is not virialized, it could give rise to local inhomogeneities, e.g. subhalos, tidal streams or unbound dark matter particles with velocities exceeding the escape velocity. It is worth mentioning that the effect of these assumptions on the astrophysical parameters and dark matter halo distributions on the results of different experiments is reduced by choosing the common values as introduced above. However, the effects can also be energy dependent, thus altering the detector response for diverse target materials. Therefore, other analysis methods are necessary to resolve these ambiguities (see section 5).

The dark matter density profile can only be indirectly observed (e.g. rotation velocities of stars); therefore, numerical simulations have been performed in order to understand the structure of halos. These simulations contained traditionally only dark matter [141–144] and showed triaxial velocity distributions [140]. The resulting haloes feature, however, cusped profiles with steeper density variations towards the centre of the halo, while observations favoured flatter cored-profiles. Moreover, the simulations predict a large amount of substructure, i.e. large number of subhaloes, in contradiction with the few haloes present in the Milky Way. These issues, currently under investigation, might challenge the validity of the ΛCDM model and different possible solutions are discussed. One solution could be related to the nature of dark matter or its properties [145]. A warm dark matter candidate with a larger free-streaming length could, for instance, modify the halo density profile resulting in the observed cored-type profiles and suppressing the formation of small structure. Another possibility is to consider candidates with weak interaction with matter but strong self-interaction [146]. The elastic scattering of these particles in the dense central region could modify the energy and momentum distribution resulting in cored dark matter profiles. Probably, the solution could be related to the absence of baryonic matter in the simulations. The effect of baryons to the halo mass distribution is observed, for instance, in the recent Illustris-1 simulation [8] which considers the coevolution of both dark and visible matter in the Universe. Furthermore, sudden mass outflows can alter substantially the central structure of haloes [147]. Dark matter simulations including also baryons [16] show how gas outflows can change the distribution of gas and stars. For sufficiently fast outflows, the dark matter distribution can be also affected explaining hereby the low central-halo densities.

Nevertheless even with large simulations containing baryons, uncertainties in the dark matter halo remain and, thus, direct-detection experiments generally use the common assumption of an isotropic Maxwell–Boltzmann distribution using values for astrophysical parameters as introduced above. In section 5.3, a method to display results in an astrophysical independent representation is described.

4.   Background sources and reduction techniques

In order to identify unambiguously interactions from dark matter particles, ultra-low background experimental conditions are required. This section summarises the various background contributions for a direct dark matter experiment. It includes external radiation by γ-rays, neutrons and neutrinos which is common for all experiments and internal backgrounds for solid-state and for liquid detectors. The main strategies to suppress these backgrounds through shielding, material selection, and reduction in data analysis are also discussed.

Environmental gamma-ray radiation

The dominant radiation from gamma-decays originates from the decays in the natural uranium and thorium chains, as well as from decays of common isotopes e.g. ^{40}K, ^{60}Co and ^{137}Cs present in the surrounding materials. The uranium (^{238}U) and thorium (^{232}Th) chains have a series of alpha and beta decays accompanied by the emission of several γ-rays with energies from tens of keV up to 2.6 MeV (highest γ-energy from the thorium chain). The interactions of γ-rays with matter include the photoelectric effect, Compton scattering and $e^- e^+$ pair production [148]. While the photoelectric effect has the highest cross-section at energies up to few hundred keV, the cross-section for pair production dominates above several MeV. For the energies in between, Compton scattering is the most probable process. All these reactions result in the emission of an electron (or electron and positron for pair production) which can deposit its energy in the target medium. Such energy depositions can be at energies of a few keV affecting the sensitivity of the experiments because this is the energy region of interest for dark matter searches.

Gamma radiation close to the sensitive volume of the detector can be reduced by selecting materials with low radioactive traces. Gamma-spectrometry using high-purity germanium detectors are a common and powerful technique to screen and select radio-pure materials. Other techniques such as mass spectrometry or neutron activation analysis are also used for this purpose [149]. The unavoidable gamma activity from natural radioactivity outside the experimental setup can be shielded by surrounding the detector by a material with a high atomic number and a high density, i.e. good stopping power, and low internal contamination. Lead is a common material used for this purpose. Large water tanks are also employed as they provide a homogeneous shielding as well as the background requirements. To reduce the γ-ray activity from radon in the air, the inner part of the detector shield is either flushed with clean nitrogen or the radon is reduced using a radon trap facility [150].

Analysis tools can be used to further reduce the rate of background interactions. Given the low probability of dark matter particles to interact, the removal of multiple simultaneous hits in the target volume can be, for instance, used for background-event suppression. This includes tagging time-coincident hits in different crystals or identifying multiple scatters in homogeneous detectors. For detectors with sensitivity to the position of the interaction, an innermost volume can be selected for the analysis (fiducial volume). As the penetration range of radiation has an exponential dependence on the distance, most interactions take place close to the surface and background is effectively suppressed. Finally, detectors able to distinguish electronic recoils from nuclear recoils (see section 5.1) can reduce the background by exploiting the corresponding separation parameter.

Cosmogenic and radiogenic neutron radiation

Neutrons can interact with nuclei in the detector target via elastic scattering producing nuclear recoils. This is a dangerous background because the type of signal is identical to that of the WIMPs. Note that there is also inelastic scattering where the nuclear recoil is typically accompanied by a gamma emission which can be used to tag these events. Cosmogenic neutrons are produced due to spallation reactions of muons on nuclei in the experimental setup or surrounding rock. These neutrons can have energies up to several GeV [151] and are moderated by the detector surrounding materials resulting in MeV energies which can produce nuclear recoils in the energy regime relevant for dark matter searches. In addition, neutrons are emitted in (α, n)- and spontaneous fission reactions from natural radioactivity (called radiogenic neutrons). These neutrons have lower energies of around a few MeV.

Dark matter experiments are typically placed at underground laboratories in order to minimize the number of produced muon-induced neutrons. The deeper the location of the experiment, the lower the muon flux. Figure 3 shows the muon flux as a function of depth for different laboratories hosting dark matter experiments.

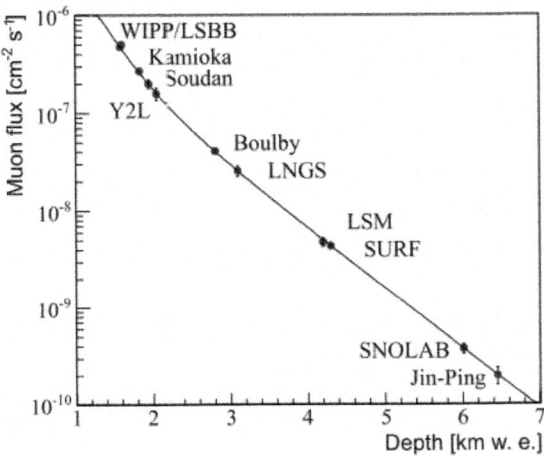

Figure 3: Muon flux as function of depth in kilometres water equivalent (km w. e.) for various underground laboratories hosting dark matter experiments. The effective depth is calculated using the parametrisation curve (thin line) from [151].

The effective depth is calculated using the parametrisation from [151] which is represented by the black line in the figure. The muon flux for each underground location is taken from the corresponding reference of the list below.

Waste Isolation Pilot Plant (WIPP) [152] in USA.

Laboratoire Souterrain à Bras Bruit (LSBB) [153] in France.

Kamioka observatory [151] in Japan.

Soudan Underground Laboratory [151] in USA.

Yang Yang Underground Lab (Y2L) [154] in Korea.

Boulby Underground Laboratory [151] in UK.

Laboratori Nazionali del Gran Sasso (LNGS) [151] in Italy.

Laboratoire Souterrain de Modane (LSM) [155] in France.

Sanford Underground Research Facility (SURF) [151] in USA.

SNOLAB [151] in Canada.

Jin-Ping laboratory [156] in China.

The flux of radiogenic neutrons can be reduced via material selection. Detector materials with low uranium and thorium content give lower α- and spontaneous fission rates. In addition, detector shielding can be used to reduce the external neutron flux further. Often water or polyethylene layers are installed around the detector setup to moderate the neutrons effectively [157]. Active vetoes are designed to record interactions of muons. The data acquired in the inner detector simultaneously to the muon event are discarded in order to reduce the muon-induced neutron background. Plastic scintillator plates are, for example, used for this purpose [150, 158]. This can be improved further by the use of water Cherenkov detectors [159, 160] as they provide a higher muon tagging efficiency (full coverage), are efficient in stopping neutrons and, for sufficiently large thickness, the external gamma activity is also reduced. To tag directly the interactions of neutrons, shielding using liquid scintillators can be used [161].

Finally, the analysis techniques described in the previous section can also be applied to reduce the neutron background. The multiple scattering tagging is, for instance, particularly effective with growing size of targets. The fiducial volume selection can also be used; however, it has a smaller effect in the reduction of background for neutrons than for gamma interactions because of the larger mean free path of neutrons.

Neutrino background

With increasing target masses approaching hundreds of kilograms to tons, direct dark matter detectors with sensitivity to keV energies start being sensitive to neutrino interactions. Neutrinos will become, therefore, a significant background contributing both to electronic and nuclear-recoils. Solar neutrinos can scatter elastically with electrons in the target via charged and neutral current interactions for ν_e and only neutral current for the other neutrino flavours [162]. Due to their larger fluxes, pp- and ^{7}Be-neutrinos would be the first neutrinos which could be detected. The resulting signal is a recoiling electron in contrast to the nuclear recoil resulting from WIMP interactions. Therefore, neutrino–electron scattering is an important background mainly for experiments which are not able to distinguish between nuclear and electronic recoils (see section 5.1). Here, we consider neutrino-induced reactions as background but the measurement is interesting on itself as it can confirm the recent pp-neutrino measurement by the Borexino experiment [163], testing in real time the main energy production mechanism inside the Sun.

Neutrinos can also undergo coherent neutrino–nucleus elastic scattering producing nuclear recoils with energies up to few keV [164]. Although this process has not been measured yet, it is expected to be accessible in the experiments planed to run in the next couple of years. Dark matter detectors could be, hereby, the first to measure this process. Coherent scattering of solar neutrinos would limit the sensitivity of dark matter experiment for low WIMP masses (few GeV) for cross-sections around ~10^{-45} cm^2. For higher WIMP masses, the coherent scattering of atmospheric neutrinos would limit dark matter searches at ~10^{-49} cm^2 [165–167] (see also figure 17). In the case of a positive signal at these cross-sections, in principle, the modulation of the signal along the year could be considered in order to distinguish WIMPs from neutrinos. While the WIMP rate should peak around 2 June (see section 3.1), the rate of solar neutrinos should peak around 3 January due to the larger solid angle during the perihelion. The rate of atmospheric neutrinos also peaks around January [168] due to the changes in atmospheric density resulting from seasonal temperature variations.

Internal and surface backgrounds

In contrast to the external background which are common to all types of detectors, internal backgrounds differ depending on the target state. Therefore, internal backgrounds for crystal and liquid targets are discussed separately.

Crystalline detectors such as germanium or scintillators are grown from high purity powders or melts. During the growth process remaining impurities are effectively rejected as their ionic radius does not necessarily match the space in the crystalline grid. In this way, the crystal growing process itself reduces internal contaminations, for instance with radium, uranium or thorium [169–172]. Important for these detectors is the surface contamination with radon decay products. Either α-, β-decays or the nuclear recoils associated to the latter can enter the crystal, depositing part of their energy. The incomplete collection of signal carriers results in events that appear close to the region of interest, where nuclear recoils from WIMP interactions are expected. To identify events happening close to the surface, new detector designs have been developed over recent years. For example, in germanium detectors interleaved electrodes can be placed on the detector surface in order to collect an additional signal identifying the position of the event [173, 174]. In scintillating crystals, an effective reduction of surface-alpha events has been achieved by a new design with a fully scintillating surface [175]. More details will be discussed in chapter 7.

Furthermore, cosmic activation of the target or detector surrounding materials during the time before the detector is placed underground needs to be considered. One of the most important processes in the production of long-lived isotopes is the spallation of nuclei by high energy protons and neutrons. As the absorption of protons in the atmosphere is very efficient, neutrons dominate the activation at the Earth's surface for energies below GeV [176]. Exposure time, height above sea level and latitude affect the yield of isotopes; therefore, by minimizing the time at surface and avoiding transportation via aircraft, the isotope creation can be reduced. Since these precautions cannot always be taken, tools or studies targeted to quantify the background due to cosmogenic activation are required (see for example [176–179]).

For noble gases, a contribution to the internal background originates from cosmogenic-activated radioactive isotopes contained in the target nuclei. For argon, ^{39}Ar with an endpoint energy at 565 keV has a large contribution as it is produced from cosmic-ray activation at a level of 1 Bqkg^{-1} in natural argon. In order to reduce it, argon from underground sources is extracted. It has been shown that in this way, the activity is reduced by at least a factor of 100 [180]. In xenon, cosmic activation produces also radioactive isotopes, all rather short-lived.^{127}Xe has the longest lifetime with 36d which is still short enough to decay within the start of the experiment [181]. Xenon also contains a double beta decaying isotope, ^{136}Xe; however, its lifetime is so large, 2.2×10^{21} y [182], that it does not contribute to the background for detectors up to few tons mass. If necessary, this isotope can be removed relatively easily by centrifugation. In addition, decays from the contamination of the target with krypton and the radon emanation from the detector materials contribute to the internal background. The β-decaying isotope ^{85}Kr is produced in nuclear fission and it is released to the atmosphere by nuclear-fuel reprocessing plants and in tests of nuclear weapons. Krypton can be removed from xenon either by cryogenic distillation [183] or using chromatographic separation [159]. Both methods have been proven to work at the XMASS/XENON and LUX experiments, respectively. Besides the reduction of krypton in the target, techniques to determine the remaining krypton contamination are necessary in order to precisely quantify its contribution the remaining contamination. Recently, detections in the ppq (parts per quadrillion) regime of natural Kr in Xe have been achieved [184]. Another possible method is the use of an atom-trap trace analysis system [185]. Radon is emanated from all detector materials containing traces of uranium or thorium. Once radon is produced in these decay chains, it slowly diffuses throughout the material and can be then dissolved in the liquid target. An approach to reduce radon is to use materials with low radon emanation [186, 187]. Furthermore, methods to continuously remove the emanated radon are being investigated [188, 189].

For both solids and liquids, the surface deserves special attention. For example, radium accumulated at the surfaces of the target or in the materials in contact with the liquid can contribute to the background, i.e. surface background and radon emanation. Surface treatment with acid cleaning and electro polishing has been proven to be effective in removing radioactive contaminants at the surfaces [190].

5.   Result of a direct detection experiment

This section gives a generic description of a dark matter experimental result starting with the signal production in the target media. The statistical treatment of the measured events is discussed as well as the representation of the derived results.

Detector signals

The elastic scattering of a dark matter particle off a target medium induces for the case of the WIMP an energy transfer to nuclei which can be observed through three different signals, depending on the detector technology in use. These can be the production of heat (phonons in a crystal), an excitation of the target nucleus which de-excites releasing scintillation photons or by the direct ionization of the target atoms. Detection strategies focus either on one of the three signals, or on a combination of two of them. Although, in principle, all three signals could be recorded, such an experiment does not exist to date. Figure 4 shows a scheme of the possible observables, as well as the most common detector technologies. A combination of two detection channels turns out to be powerful, since the response of media to an interaction is not only proportional to the deposited energy but depends on the type of particle that deposits the energy. More precisely, the relative size of the two signals depends on the type of particle. This enables the discrimination of nuclear recoils (e.g. neutrons, WIMPs) from electronic recoils (e.g. photon interactions, beta decays) which is an important method to reduce the background of the experiment. To measure the ionization signal either germanium detectors or gases (low pressure, for directional searches) are employed while scintillation can be recorded for crystals and for noble-gas liquids. To detect heat, the phonons produced in crystals are collected using cryogenic bolometers at mK temperatures. The heat signal is also responsible for nucleation processes in experiments using superheated fluids. Detectors which explore the discrimination power by measuring two signals are positioned in figure 4 between the corresponding signals: scintillating bolometers for phonon and light detection, germanium or silicon crystals to measure phonon and charge, and double phase (gas–liquid) noble-gas detectors for charge and light read-out. We mention that discrimination can also be achieved by exploiting other features in the response of the medium. For instance, the pulse shape of the signal depends on the particle type in liquid noble-gas scintillators. Detailed information on the various detector technologies used in direct dark matter searches is given in section 7.

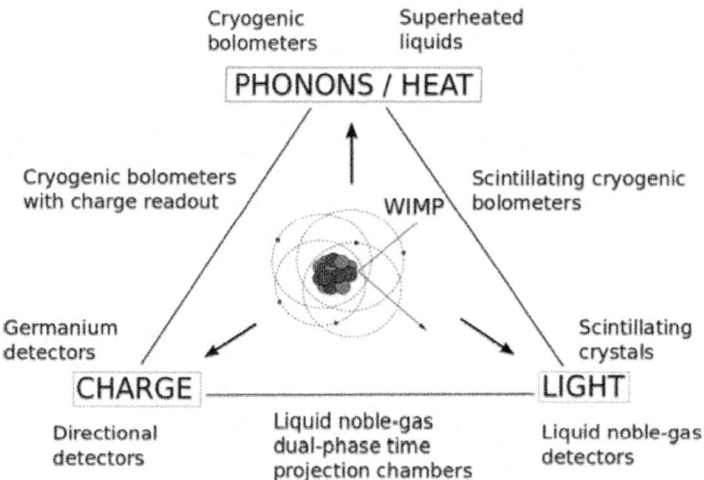

Figure 4: Schematic of possible signals that can be measured in direct-detection experiments depending on the technology in use.

Statistical treatment of data

In direct-detection experiments, various statistical methods are used to derive upper limits on the WIMP–nucleus cross-section as a function of the dark matter mass or to claim a detection of dark matter. Over recent years, a number of experiments have recorded events above the expected background and based on those, signal contours in cross-section with nucleons versus dark matter mass have been derived [191–194]. Some of those results have been, later on, disfavoured by the same authors based on new data from upgraded detectors. In this potential 'discovery' situation, a correct application of statistical methods is essential to avoid a misidentification of up- or downward fluctuations of the background. Common to all experiments is not only that the expected signal consists of only a few events per year but also an unavoidable presence of background (see section 4 for a through explanation). Hence, a statistical analysis has to consider both the Poisson distribution of the signal events and a correct treatment of systematic uncertainties of the detector response. A detailed description of the methods can be found in [195].

For detectors featuring a separation between different types of particles (discrimination), an intuitive approach is to select a signal region where the ratio of signal to expected background is high. This is indicated by the blue rectangle in the left panel of figure 5. Due to the generally low number of expected events, their expectation value is described by a Poisson distribution. If the knowledge of the background distribution is available, e.g. using calibration data or a Monte Carlo (MC) simulation, a background prediction for the signal region can be estimated. An exclusion limit (one-sided confidence interval) or an interval representing the uncertainties on a possible signal (two-sided confidence interval) can be computed based on a likelihood ratio of Poisson distributions developed by Feldman and Cousins [196]. This method gives the correct coverage, i.e. a quantification of how often the interval contains the true value of interest, and is able to decide between one- and two-sided confidence intervals. However, the knowledge of the background and signal probability density functions cannot be exploited, nor can an uncertainty in the background prediction be addressed. It is worth mentioning that other methods exist which include systematic uncertainties in the confidence interval construction [197, 198]. The Feldman and Cousins method is used, for example, for the derivation of results of the PandaX [199], ZEPLIN-III [200] and SIMPLE [201] experiments.

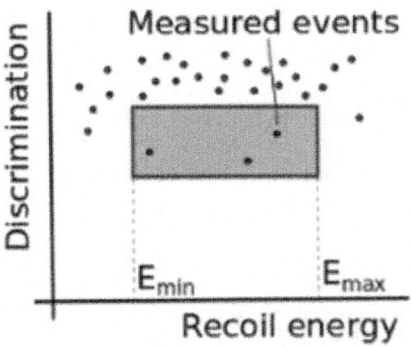

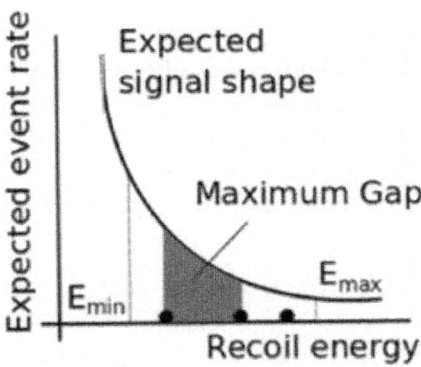

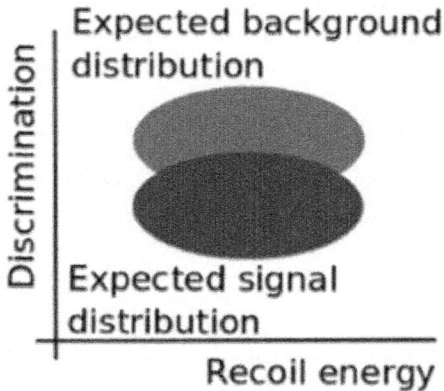

Figure 5: The left figure is an illustration of the statistical analysis method where the result is reduced to a simple counting experiment in the Poisson regime in the presence of background (e.g. Feldman and Cousins [196]). Yellin's method improved towards the simple box-based analysis by further considering the signal shape to derive limits (middle figure). If the probability density functions for the background and signal distribution are known, limits as well as a discovery can be calculated with a maximum likelihood analysis (right). For more information see text.

If no background prediction is available or possible, or its uncertainty is too large, Yellin's method [202] can be used. In this method, the absence or low density of events in certain energy intervals (maximum gap or optimal interval, respectively) is used to calculate the probability of not measuring dark matter events in that interval. The middle panel of figure 5 shows how the maximum gap (red area) is defined given the measured data (black points) and the knowledge on the signal shape (black curve). By not assuming a specific background model, the method is robust against unexpected background events. A one-sided interval (upper limit) in the presence of unknown background can be calculated leading always to conservative results which take into account the signal shape. By construction this method leads to one-sided confidence intervals and, therefore, no signal discovery is possible. Current published upper limits derived by this method are e.g. Super CDMS [203], CRESST-II [204] and PICO [205].

Exploiting the full knowledge of the signal as well as the background distributions, allows one to use a maximum likelihood estimation which typically results in stronger exclusion limits or a higher significance of a signal (right panel in figure 5). Furthermore, nuisance parameters can be treated in the context of a profile likelihood analysis in order to account for systematic uncertainties. These parameters are not of immediate interest for the analysis of the signal model and, therefore, their uncertainties can be profiled out

[206, 207]. This method not only allows one to penalize the result due to a limited knowledge of the detector response but also enables a natural transition between one- and two-sided confidence intervals. For setting exclusion limits, the maximum or profile likelihood analysis might be combined with a method developed in [208, 209] to reduce the impact of a statistical downward fluctuation of the background on the result. The results of XENON [210], LUX [211] and CDMS II Ge [212] use this method to derive upper limits while in CDMS II Si [193] it was used to calculate the significance of the measured events above the expected background.

Experimental results can be computed either by a frequentist or Bayesian interpretation of the data. The former is extensively employed for direct-detection experiments as reviewed above. The latter is, so far, less common in the analysis of dark matter experiments. The Bayesian interpretation of likelihood and probability differs from the frequentist approach. In contrast, to state the frequency of possible outcomes of an experiment by confidence intervals, Bayesian credible intervals allow a statement about the degree of belief of the tested hypothesis and are based on the Bayes' theorem. Thus, the computation of a probability for a theoretical model to be true based on the observed data is only possible with Bayesian statistics. In addition, it is necessary (or possible) to assign a priori information in the form of a prior which might bias the result if not chosen appropriately. Systematic uncertainties can, similarly to the profile likelihood method, be considered in the Bayesian framework which are later marginalised by Monte Carlo Markov chains (MCMCs) e.g. the Metropolis–Hastings algorithm [213, 214]. A more detailed review of a Bayesian analysis of direct-detection experiments can be found in [215].

Generic result of a direct detection experiment

The data of a dark matter experiment are an event rate consisting usually of only a few counts and featuring a certain spectral shape (see section 3.1). These results are commonly displayed in a parameter space of the dark matter–nucleon cross-section and the dark matter mass. To derive these physical properties, astrophysical values for the dark matter density and its velocity distribution have to be assumed (see section 3.4).

The most common way to display direct-detection results is based on a differential rate with SI and isospin-conserving interactions

$$\frac{dR}{dE}(E, t) = \frac{\rho_0}{2\mu_A^2 \cdot m_\chi} \cdot \sigma_0 \cdot A^2 \cdot F^2 \int_{v_{min}}^{v_{esc}} \frac{f(\mathbf{v}, t)}{v} \, d^3v,$$

(13)

with v_{esc} the escape velocity (see section 3.4) and the minimal velocity defined as

$$v_{min} = \sqrt{\frac{2m_A \cdot E_{thr}}{2\mu_A^2}} .$$

(14)

The parameter E_{thr} describes the energy threshold of the detector and μ_A is the reduced mass of the WIMP–nucleus system. The left plot in figure 6 shows a generic limit (open black curve) on the dark matter cross-section with respect to the dark matter mass which can be calculated with equation (13). At low WIMP masses the sensitivity is reduced mainly due to the low-energy threshold of the detector, whereas the minimum of the exclusion curve is given by the kinematics of the scattering process which depends on the target nucleus. At larger WIMP masses, the event rate is overall suppressed by $1/m_\chi$. Given that the local dark matter density is a constant of 0.3 GeV cm^{-3} (section 3.4), the heavier the individual particles, the fewer particles are available for scattering. In addition, the form factor reduces the rate for interactions with a large momentum transfer (section 3.2). The overall sensitivity of the experiment is dominated by the product of the size of the target and duration of the measurement, also called exposure, as well as the ability to avoid or reduce background events. A possible detection would be displayed as a contour region (closed black line) representing a certain confidence level. The coloured lines indicate qualitatively the influence of varying certain detector parameters. The exposure can be increased by either longer measurements or by an increased target mass. An increase in exposure enhances the ability to measure lower cross-sections (green line). Note, however, that typically the background scales up with larger target masses and reduces the sensitivity if it is not simultaneously suppressed with improved techniques. By using lighter target nuclei (red line), not only is the kinematics of the elastic scattering modified but also v_{min} is reduced, resulting in a shift of the maximum sensitivity to lighter WIMP masses. The event rate is proportional to A^2 and thus a smaller value of A reduces the overall sensitivity. Lowering the energy threshold of the detector (blue line) allows one not only to extend the sensitivity to lighter WIMP masses but also reduces the value of v_{min} and, hence, allows one to test smaller cross-sections.

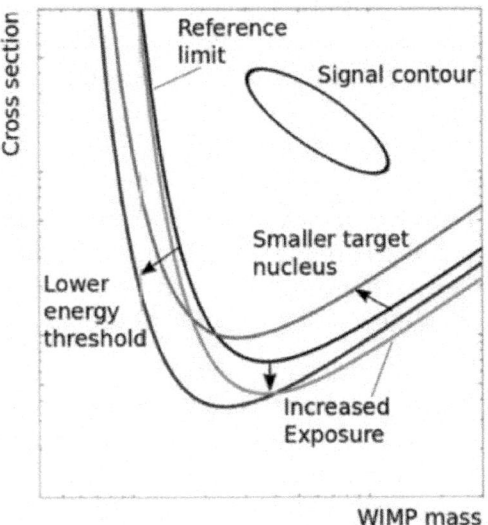

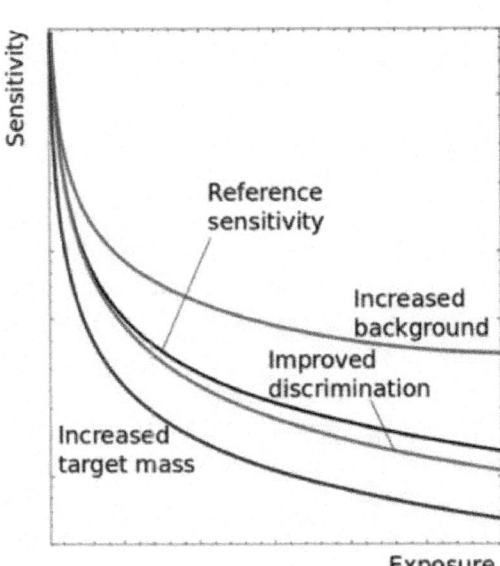

Figure 6: Left: illustration of a result from a direct dark matter detector derived as a cross-section with matter as function of the WIMP mass. The black line shows a limit and signal for reference, while the colored limits illustrate the variation of an upper limit due to changes in the detector design or properties. Right: evolution of the sensitivity versus the exposure. For more information see text.

The right plot in figure 6 illustrates the evolution of the sensitivity to the cross-section with respect to the exposure. For a given detector mass, the increase in exposure is caused by the accumulation of measuring time. The black line shows a reference curve. A non-discriminating detector, or a discriminating detector with an order of magnitude higher background, reaches its maximum sensitivity sooner (red line), and longer measurements do not improve the sensitivity. Assuming a constant background while enlarging the target mass, the sensitivity still increases with time (blue) and is not yet statistically limited. As mentioned before, to keep a low background level requires a higher purity of the detector and target material. An improved discrimination between background and signal events improves the acceptance to signal events and can also lead to a higher sensitivity (green line).

The choice of a different dark matter halo model, $f(\mathbf{v}, t)$, affects the comparability of results from experiments using different target materials and technologies, as they might probe different dark matter velocity intervals [216]. Therefore, an alternative representation of the data which integrates out astrophysical uncertainties has been proposed [217, 218]. By displaying results in a parameter space which is halo independent, direct comparisons of detectors are feasible. This is possible by defining a parameter η containing all astrophysical assumptions

$$\eta = \frac{\rho_0 \cdot \sigma_0}{m_\chi} \cdot \int_{v_{\min}} \frac{f(\mathbf{v}, t)}{v} \mathrm{d}^3 v.$$

(15)

Using the monotonicity of the velocity integral [217], η can be approximated to be independent of the detector response function and, hence, to be common to all experiments. Figure 7 shows examples of exclusion limits (black and green lines) and signals (blue crosses) for a fixed dark matter mass, using η and the dark matter velocity as free parameters. The dark matter velocity is defined as the minimum speed that a WIMP requires in order to deposit a certain nuclear-recoil energy in the detector. The red line represents a typical halo model and all velocities below the line describe its feasible physical parameters. Experiments can probe the minimal velocity region above their exclusion limits up to the halo model (red line). Due to the use of different target elements and energy thresholds, these regions differ for each experiment. Hence, the limits from A and B cannot be compared to each other since they probe different velocity intervals of the halo model. Accordingly, only one- and two-sided confidence intervals within the same velocity interval are robust against the uncertainties of astrophysical parameters. This parameter space has been used in the last couple of years to display results of direct detection experiments (see as examples [219–222]).

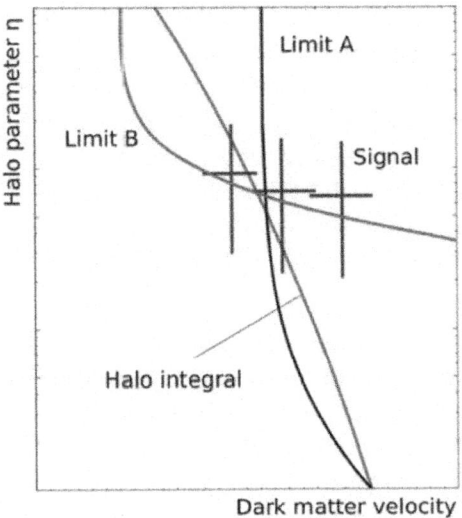

Figure 7: Illustration of the result of a direct-detection experiment in the parameter space η(see text) which is free of astrophysical assumptions and allows a direct comparison of different experiments. The blue markers indicate a dark matter signal, whereas the black and green lines indicate exclusion limits. The red line shows the velocity integral for a fixed choice of halo parameters.

DETECTOR CALIBRATION

Results from dark matter detectors contain typically a low number of signal-like interactions for which their recoil energies are measured. In order to understand and interpret these data, the energy scale for the recoiling nucleus has to be characterized. This is particularly important to determine the energy threshold and to possibly constrain the WIMP mass. In addition, the signal and background regions have to be calibrated in the parameter space relevant for the analysis. For this purpose, regular calibrations using, for instance, radioactive sources are carried out. This section summarises the strategies and main features of calibration for different detector technologies focusing on the most competitive detector types.

Calibration of the Recoil-Energies

Depending on the detector technology (see section 7), scintillation photons, phonons in a crystal and/or charge signal from ionization can be measured. For a given energy deposition after a particle interaction, the corresponding recoil energy can be calculated applying a conversion function which contains quenching effects, i.e. the losses of signal due to various mechanisms as function of recoil energy. This function is, in general, different for electronic recoils and for nuclear recoils as the quenching mechanisms depend on the energy and nature of the interacting particle. In order to emphasize which function has been used for the conversion, energy units are expressed in electronic-recoil equivalent keV_{ee} or nuclear-recoil equivalent keV_{nr}. Experiments recording phonons in a crystal lattice collect the full recoil energy in the form of phonons [223–225] and therefore, no signal quenching is usually considered. However, inhomogeneities inside the crystal (e.g. crystal defects) can, in principle, lead to phonon quenching.

WIMPs are assumed to produce nuclear recoils and, hence, an energy scale to convert the measured quanta to nuclear-recoil equivalent energy (keV_{nr}) is required. This scale is determined either with direct measurements using neutron scattering experiments or by comparing the nuclear recoil spectra of calibration neutron-sources to MC generated ones. The first method is widely used as it is, in general, more robust due to the fewer assumptions involved compared to MC methods. In direct experiments (see scheme in figure 8), mono-energetic neutrons are scattered once in the medium and once in a detector operated in coincidence. By choosing a scattering angle and selecting single interactions, the kinematics are determined and a mono-energetic nuclear recoil can be selected. Varying the scattering angle provides the energy dependence of the signal yield. The results are usually normalized to the yield of a known electronic energy deposition. This method tests directly the kinematics of the WIMP interaction since single-scattering neutron events are selected similarly to the expected WIMP scattering process.

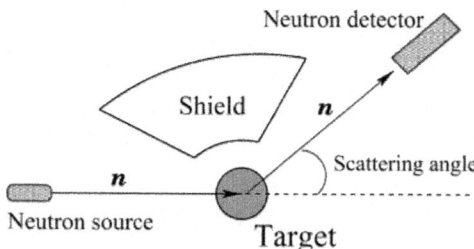

Figure 8: Scheme of a neutron scattering experiment using a neutron source, a target material under study and a coincidence neutron detector to determine the response of the target to mono-energetic nuclear recoils.

Figure 9 shows the nuclear-recoil energy dependence of the signal yield for various media at energies relevant for dark matter searches. The top left figure shows the ionization efficiency for germanium using the data provided by the authors of [226] and references therein. The measurements are performed down to energies about 1 keV$_{nr}$. The light yield quenching is shown for sodium in a NaI crystal (top right), for liquid xenon (LXe) (bottom left) and liquid argon (LAr) (bottom right). The data are from [227–230] including references therein for Na, LXe and LAr, respectively. It can be seen that for sodium and xenon several measurements exist with some of them exploiting the quenching down to a couple of keV$_{hr}$ energies. The most recent experimental results show a decrease of efficiency at decreasing energies. Older measurements have increasing efficiency with large error bars which could be due to an incomplete consideration of the detection efficiency [231]. For argon, there exist two sets of recent measurements by [230, 232] both with relatively small errors. These two results are in contradiction to each other within several sigma and further investigations will be necessary to understand the scintillation behaviour at low recoil energies. For LAr and LXe detectors operated with an electric field (see section 7.4), the field quenching has to be considered. For argon, dedicated measurements have shown a significant dependence of the light yield on the field, up to 32% for energies between 11 and 50 keV$_{nr}$ [233]. Note that this energy scale is used to derive the energy threshold of the experiment in keV$_{nr}$. As the WIMP recoil spectrum has an exponential shape (see section 5), uncertainties in the threshold result in large variation of the expected number of events, especially at low WIMP masses. Therefore, usually the energy threshold of an experiment should be in a region for which quenching data points exist.

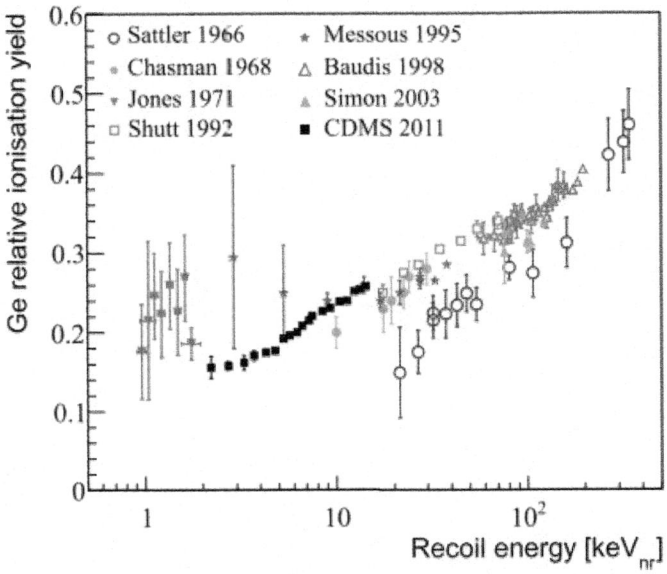

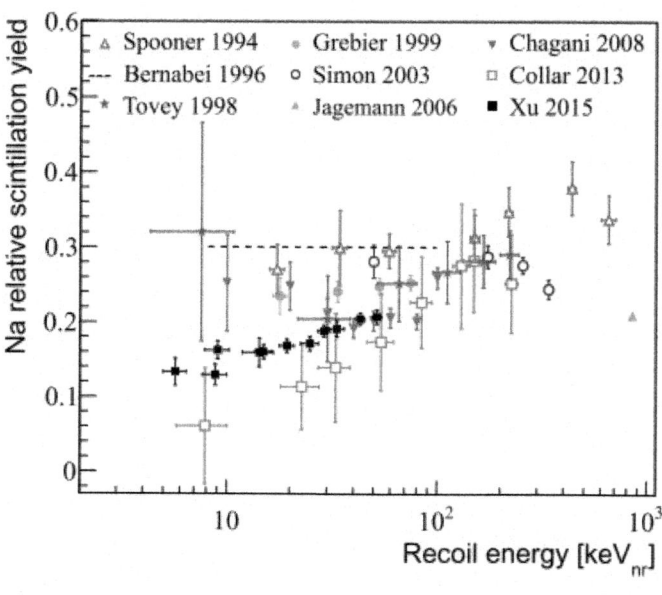

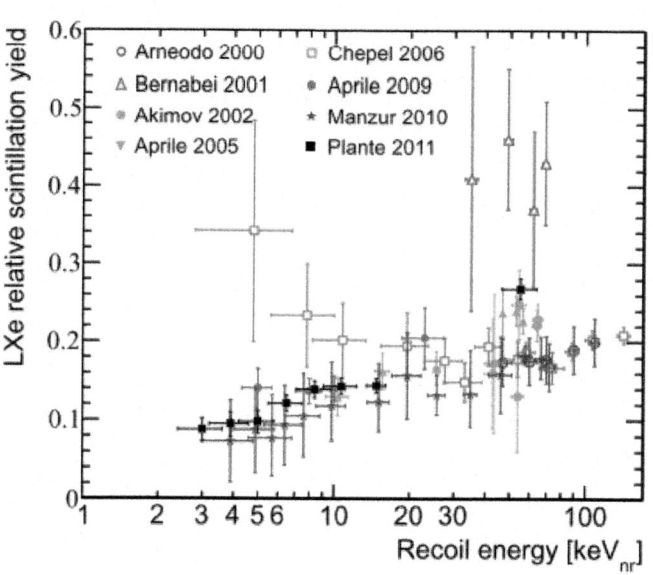

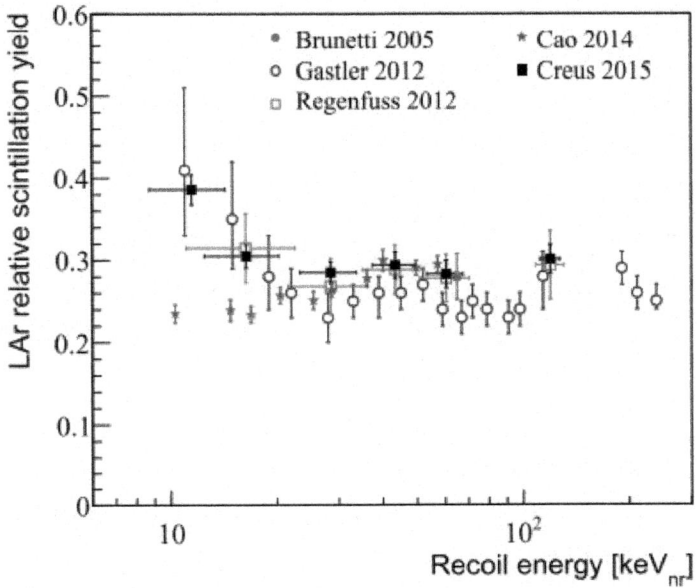

Figure 9: Measurements of signal quenching for various detector materials: ionization efficiency in germanium (top left, from [226] and references therein), scintillation efficiency for Na (top right, from [227] and references therein), for LXe (bottom left, from [228] and references therein) and for argon (bottom right, from [229, 230] and references therein).

A second method to derive the energy scale of a detection medium is to compare the spectrum of a neutron calibration-source to an MC generated one. This method has been used for example for LXe detectors [234, 235] giving results which are compatible with direct measurements. A complementary approach to describe the signal yield as a function of recoil energy is the theoretical modelling of the underlying processes; however, the accuracy of such descriptions has to be tested with data. Often the model proposed by Lindhard [236] is adopted but several other models exist. Some of them try to describe the electronic and nuclear stopping power at low energies as in [226] for ionization in germanium. This model is verified by a dedicated data-MC comparison of neutron scattering off germanium [237]. In [238–240], the scintillation and ionization for a liquid noble-gas detector is modelled. The former includes scintillation quenching from the Birks [241] model while the latter two incorporate measured data into the description.

Note that for superheated liquid detectors, the energy calibration differs from the experiments mentioned above. The nuclear recoil scale is calculated using the 'hot spike' model of bubble nucleation [242] and it is verified by experimental data. For example, the three consecutive α-decays from ^{222}Rn can be used to test the model [243]. By varying the pressure and temperature of the detector volume, the energy threshold for a bubble nucleation is chosen. For different threshold energies, number of events is measured without the determination of the individual recoil energies (see also section 7.5).

In this section, the energy scale relevant for nuclear recoils produced by a WIMP-like dark matter candidate has been discussed. However, for candidates interacting with electrons instead of with the nucleus, the corresponding electronic-recoil energy has to be applied. This scale is measured using mono-energetic signals from photo-/full-absorption of γ-rays, or by Compton effect coincidence experiments (see for example [244]) similar to the neutron-scattering mentioned above.

Determination of signal and background regions

Most experiments searching for WIMP interactions in a target material use either the combination of two signals (phonon, light or charge) or the pulse-shape of the signal to distinguish between the main background from electronic recoils by γ- and β-decays from the nuclear recoil signal. The signal and background regions are typically defined via dedicated calibration campaigns in between the science data taking. The distribution of nuclear recoils can be studied selecting interactions of neutron sources such as ^{241}AmBe or ^{252}Cf. It is important to acquire enough nuclear-recoil statistics to have a precise determination of the signal region. In addition, the signal acceptance has to be quantified since this quantity enters directly into the sensitivity of the experiment.

The modelling of the background composition of each experiment is required to calibrate the various components adequately. As most of the background arises from electronic recoils fromγ-interactions in the target, the background region can be determined by exposing the detector to gamma sources at different positions. Commonly, radioactive sources like ^{133}Ba, ^{137}Cs, ^{60}Co or ^{232}Th are used. For liquid noble-gas detectors internal background also contributes and internal sources can be used to characterize them (e.g. a tritiated source used by the LUX experiment [211]). For solid-state detectors surface events also need to be characterized (see section 4.4). This is typically carried out by exposing the crystals to β- or α-emitters at different locations on the surface of the detector. Also in this case, it is desired to perform a high statistics measurement of the background as it enters the background prediction and its uncertainty. For superheated liquid detectors, the thermodynamic conditions are adjusted such that the medium is not sensitive to γ-rays or electrons.

Therefore, only the background from α-decays needs to be characterized (see section 7.5).

Figure 10 shows schematically how signal (in blue) and background (in red) events are distributed for some detector technologies. On the left the ionization yield of a germanium bolometer is represented: the phonon signal is used to determine the energy scale and the normalized ratio of phonon to ionization signal is used for signal discrimination. This type of detector achieves with this method a large separation of signal and background, e.g. a 10^6 rejection of electronic recoils can be achieved [245].

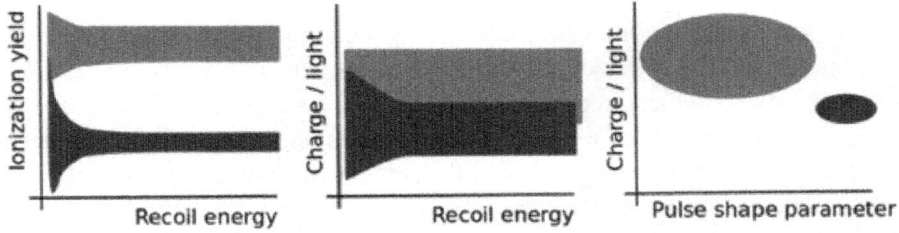

Figure 10: Schematic representation of signal (blue) and background (red) regions for a bolometer like a germanium detector (left), a liquid xenon TPC (middle) and a liquid argon TPC (right).

Only surface events with incomplete charge collection produce events leaking from the background region down to the signal region. The middle panel of figure 10 shows the separation of signal and background for a LXe detector using the ratio of charge to light signals. The signal discrimination is typically not as large as for bolometers. The highest γ-ray rejection factor achieved to date is at 5×10^3 [246]. Finally, the right panel shows, for a certain energy interval, the combination of two discrimination parameters as can be done in a LAr time-projection-chamber (TPC). In addition to the charge to light signal-ratio, the pulse shape of the scintillation can be used to separate signal and background. The WARP experiment, for example, made use of these two discrimination parameters in the data analysis [247]. For each energy interval, the regions can be determined gaining a large signal acceptance and strong background suppression.

TECHNOLOGIES AND EXPERIMENTAL RESULTS

In this section, the working principle of various technologies searching for dark matter is reviewed. As mentioned in section 5, most experiments exploit either the phonon, charge or light signal, or a combination of those. In solid targets, phonon excitations of the crystal arise by the conversion of the kinetic energy from the scattering particle to lattice vibrations. The typical energy scale to create phonons in crystals is of the order of a few meV which is considerably lower than the energy of the quanta of light or charge. Charged particles moving through a medium ionize its atoms and the produced charges can be collected if an electric field is applied. To create an electron–hole pair in a semiconductor, a typical energy of a few eV is necessary [148], whereas for liquid noble gases the ionization energy is of the order of (10–20) eV [248, 249]. The photons emitted by scintillating materials are produced mainly by a relaxation of the excited medium. It is common to all scintillators (solid or liquids) that only a small fraction (1%–10%) of the total recoil energy is transferred to scintillation processes.

When describing the various existing detector technologies, the main challenges in direct detection will be considered: a low energy threshold to detect the smallest recoil energies, a low background to increase the signal significance and a large detector mass to increase the interaction probability inside the target. A fourth and maybe underestimated goal is a stable detector performance over time scales of a few years, where simpler detector configurations might be of advantage. In addition, the discrimination capabilities of different detectors will be discussed. While this section describes the main technologies used in dark matter searches, including the respective main scientific results, overview figures summarising various experimental results are shown in section 8.

Scintillator crystals at room temperature

Scintillators are some of the most used detection devices in particle physics. When radiation passes through a scintillating material, the atoms or molecules of the medium are excited and the subsequent de-excitation causes the emission of light. Among the various existing scintillators, mostly NaI(Tl) but also CsI(Tl) crystals are used in dark matter searches. In inorganic crystals, inhomogeneities are added to the crystalline structure as activators. These activators create crystal defects which act as additional luminescence centres [148]. When adding the activator thallium to NaI or CsI, the light emission of the crystals increases and the wavelength of the emitted light is shifted compared to the wavelength of the pure crystals to larger values (415 nm and 580 nm, respectively). At these wavelengths photosensors have a higher detection efficiency and the crystals show a better transparency. The advantage of these inorganic crystals is the large stopping power arising from the high density (3.7 and 4.5 g cm^{-3} for NaI and CsI, respectively) and the large light output that results in a better energy resolution (around 8% for 1 MeV energy deposition) and lower energy threshold than other scintillators. Crystals can be grown with sizes of several cm^3. Therefore, in order to achieve larger target masses, the detectors are composed of several crystals. An important advantage of this technology is its relative simplicity which allows one to operate the detectors over large time periods of several years. In these crystals only the scintillation signal is acquired, thus no particle discrimination is possible, besides the rejection of multiple hits in different crystals. An event-by-event separation of signal and background is not possible but the annual modulation of the signal (see section 3.1) can be used to identify dark matter interactions. Despite the absence of background rejection via discrimination, a high sensitivity can be achieved by keeping the overall background of the experiment sufficiently low. For this purpose, powders with low radioactive content on uranium, thorium and potassium are used to grow the crystals (see [171] as an example and section 4.4). In addition, most experiments use active vetoes operated in coincidence with the crystals to reduce further the background.

The DAMA experiment at the LNGS underground laboratory2 is searching for dark matter using ultra low-radioactive NaI(Tl) crystals [250]. A combined dataset of DAMA and its successor DAMA/LIBRA has collected 1.33 ton × y exposure showing an annual-modulated single-hit rate in the energy range (2–6) keV$_{ee}$ (keV electronic recoil equivalent, see section6.1). Its maximum is compatible with 2 June within 2σ which is the phenomenological expectation of the phase for dark matter interactions [99]. Meanwhile, the significance of this signal is at 9.3σ over a measurement of 14 annual cycles [251]. The DAMA experiment has demonstrated, hereby, that this technology allows for a stable long-term operation. Figure 11shows the residual distribution of events of the DAMA experiment as function of time together with a fit to the data (black line). The residuals are calculated from the single-hit event rate after subtracting the constant background rate.

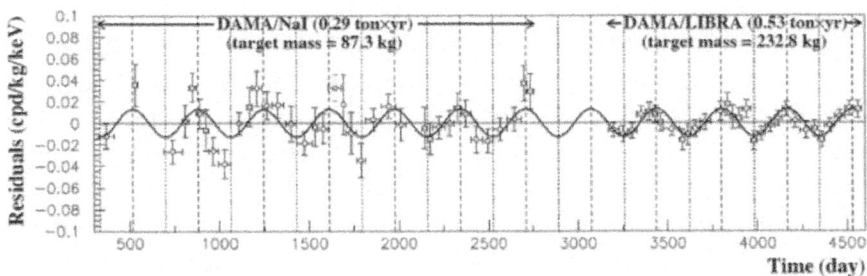

Figure 11: Annual modulation of the measured residual single-hit event rate by the DAMA experiment in the (2–6) keV energy range. The superimposed curve is a sinusoidal function with a period of one year and a phase equal to 152.5 d (maximum on 2 June). The figure covers the period between 1996 and 2007. Figure from [252].

The DAMA experiment continues taking data and since 2010, the detector has been equipped with new photosensors which will allow for a lower energy-threshold of the experiment [253].

If the DAMA signal is interpreted as being caused by elastic WIMP–nucleus interactions, two favoured regions appear at (10–15) GeV/c^2 for scattering off sodium and (60–100) GeV/c^2 for scattering off iodine [254]. The calculation of the dark matter masses depend, as pointed out by the authors of [227], on the used scintillation efficiency. DAMA measured a scintillation efficiency of 0.3 independent of the recoil energy; however, recent results shown in section 6.1 favour significantly lower efficiencies. The new determination of the efficiency would result in larger reconstructed WIMP masses for the DAMA signal. Besides the standard SI interpretation of the DAMA results, different interpretations of the annual modulation signal have been derived such as, for example, the SD [255] and the inelastic scattering of WIMPs [256] (see section 3.3). Since DAMA does not discriminate between electronic and nuclear recoil signals, dark matter candidates interacting dominantly with electrons are also considered [126].

Results from other experiments that will be presented in the following are in tension with the various dark matter interpretations of this signal. Therefore, other non-dark matter related explanations of the DAMA signal are being discussed [257]. The signal could be related to atmospheric muons, the rate of which is annually modulated due to temperature variations in the stratosphere [258], or to combinations of muons and modulated neutrinos, caused by the varying Sun–Earth distance [259]. Furthermore, varying rates of background neutrons have been considered [260]. Some of those proposals have been refuted [261, 262] but the signal and its interpretation remains controversial. A more detailed discussion about these explanations can be found in [263].

To make independent cross-checks of the DAMA signal, a number of experiments are carried out using a similar technology at various underground locations. The SABRE collaboration has proposed [171] to test the DAMA signal at LNGS, in the same underground laboratory. The experiment will consist of highly pure NaI(Tl) crystals in an active liquid scintillator veto to tag and reduce the ^{40}K background from the crystals and the external background. The Anais and the DM-Ice experiments use the same target as DAMA but are located at different locations. While a signal confirmation at the same laboratory is desired to exclude experimental effects, a measurement at a different laboratory could give information on the possible origin of the modulation. The Anais experiment [264] is currently operating a 25 kg NaI(Tl) detector at the LSC laboratory in Spain. First results showed the possibility to achieve a very low threshold at or below 2 keV$_{ee}$ due to the excellent light yield of the crystals. On the longer term, Anais aims to increase the total mass to 250 kg improving the energy threshold and the internal radioactive contamination compared to the DAMA experiment. The DM-Ice experiment is currently operating 17 kg of NaI crystals under the ice at the South Pole at a depth of 2460 m, and first results were released in 2014 [265]. This data demonstrates the feasibility of remote operation, stable environmental conditions and a background consistent with the expectation. Since the experiment is located in the Southern Hemisphere, any modulation related to seasonal effects (e.g. the modulation of atmospheric muons) would have a reverse phase to the northern hemisphere. Finally, the KIMs experiment located at the Yang Yang laboratory in Korea used an array of 103 kg CsI(Tl) crystals to test the DAMA signal caused by WIMP scattering off iodine. The energy region chosen for the analysis of the first data ranged from (3–11) keV$_{ee}$. These data, acquired between September 2009 and August 2010, showed no significant signal [266] and therefore, exclusion limits on the dark matter cross-section were derived assuming SI interactions. These limits disfavour the WIMP–iodine nuclei interactions as the source of the DAMA signal. To test the interaction on sodium, a program to develop ultra-low-background NaI(Tl) crystal detectors with lower background level and higher light yield than those of the DAMA experiment has been started [267].

Germanium Detectors

Germanium detectors combine a high radio-purity of the target material with a very low threshold down to ~0.5 keV$_{ee}$ allowing to search for WIMPs down to masses of a few GeV/c^2. Such low energies are achieved when the detectors are operated in ionization-mode, having no possibility to discriminate signal from background-like events. To reduce the noise levels sufficiently, the detectors are cooled down to the temperature of liquid nitrogen (77 K), which is, in comparison to other technologies (see section 7.3), relatively simple and does not require complex cooling systems. The noise level scales up with larger crystal sizes due to a generally increased capacitance and dedicated optimizations in the detector layout are essential [268]. The excellent energy resolution of these detectors (typically around 0.15% at 1.3 MeV) allows one to identify and quantify background sources and, eventually, this knowledge can be used to reduce the background from radioactive contaminations. In contrast to n-type doped detectors, p-type semiconductors benefit from a dead-layer around the crystal which further shields external α and β backgrounds. In addition, the rise-time of the signal can be used to discriminate surface background from bulk events. Still, a separation between electronic and nuclear recoils is not possible. At the energy threshold, the limiting feature for germanium detectors is, in general, noise from the detector itself as well as from the read-out electronics.

An ultra-low background germanium detector was already used in 1987 to derive the first limits on dark matter interactions [269]. Nowadays, this technology is further improved, particularly to reduce the energy threshold and the background level. For instance, the CoGeNT experiment [194] uses p-type point contact germanium detectors with a mass of 443 g reaching an energy threshold of 500 eV_{ee}. The CoGeNT detector has acquired, in total, 3.4 years of dark matter data in the Soudan Underground Laboratory (see section 4.2) enabling a search for dark matter by an annual modulation of the measured event rate [270]. An annual modulation of the rate was found in an energy interval of (0.5–2) keV_{ee} with a phase corresponding to the phenomenological expectation for WIMPs at a level of 2.2 σ. The amplitude of the signal is, however, a factor of 4–7 larger than expected [194]. If this signal is interpreted as SI interactions of WIMPs, a best fit value appears at a cross-section around 2.5×10^{-41} cm^2 for a 8 GeV WIMP mass. However, it should be mentioned that independent analyses on the released public data, with different assumptions for the background model [271, 272], did not find a significant signal. The CoGeNT data has also been investigated in order to search for signatures of axion-like particles [131] (see also section 2.2). These particles could interact in germanium via the axio-electric effect [273] (similar to the photo-electric effect) producing an electronic recoil of an energy corresponding to the mass of the axion. The non-observation of such a peak in the spectrum has allowed derivation of the limits on the axion–electron coupling [274]. For the future, a larger experiment, named C-4, with ten times more mass and lower background, is currently being designed [275].

A similar detector technology is used at the MAJORANA low-background broad energy germanium detector, MALBEK, operated at the Kimballton underground research facility [276]. The main motivation of this prototype detector is the demonstration of an ultra-low background level of approximately 3 events/(ton y) in the neutrino-less double beta decay ($0\nu\beta\beta$) energy region of interest. A customised 465 g germanium crystal developed specifically for a low energy threshold of 600 eV_{ee} has been tested. So far, an exposure of 89.5 kg d could be achieved, reaching a sensitivity down to 10^{-40} cm^2. This excludes part of the CoGeNT signal region using a detector with the same target material.

To avoid a high capacitance due to large crystal sizes which show generally a higher noise level, the CDEX-0 [277] experiment uses an array of four smaller (5 g) n-type germanium diodes, reaching a total exposure of 0.78 kg d. This low-threshold development is based on the detectors used by the former TEXONO experiment [278, 279] which was operated at a shallow site at the Khuo-Sheng reactor neutrino laboratory. The crystal array is surrounded by a NaI(Tl) crystal scintillator, serving as an anti-coincidence detector. The exceptional energy threshold of 177 eV_{ee} at 50% signal efficiency allows dark matter searches for SI interactions down to 2 GeV. The measured spectrum agrees well with the background expectation allowing one to place limits on the dark matter interactions. Earlier measurements with the CDEX-1 setup [280] already disfavoured the CoGeNT result. The goal for the future is to reduce the threshold to 100 eV_{ee} by increasing the total mass of the detectors.

Cryogenic Bolometers

Detectors collecting the phonon signal produced in a crystal are developed to reach very low thresholds and excellent energy resolution. If, in addition, the scintillation or charge signal is recorded, the energy dependence of the signal quenching can be used to discriminate between nuclear and electronic recoils. This can be achieved in cryogenic bolometers, where an energy deposition by a nuclear or an electronic recoil is dissipated via collisions with the nuclei and electrons in the crystal lattice. A schematic representation of the phonon-detector working principle is shown in figure 12. The dissipated energy produces phonons which can be categorized as thermal and non-thermal phonons (also called athermal). Thermal phonons are related to the thermal equilibrium of the medium after an energy deposition and can be measured by the induced temperature rise. Athermal phonons describe a fraction of the initially produced phonons which are out of equilibrium and show a larger mean free path in the medium. These phonons carry not only the information about the energy deposition but can also be used to estimate, e.g., the location of the recoil. If an electric field is applied to the crystal, e.g. to read out the charge signal, the drifted electron–hole pairs dissipate further energy in the crystal lattice producing additional phonons (Neganov–Luke effect [281, 282]). These Neganov–Luke phonons enhance the phonon signal which need to be accounted for in the estimation of the recoil energy. However, by a dedicated usage of this enhancement, the energy threshold might be significantly reduced. A general feature of cryogenic bolometers is, similar to germanium diodes, their limited crystal size (~1 kg). To achieve large exposures, these experiments generally use detector arrays which complicate not only the set up but also the analysis of the data.

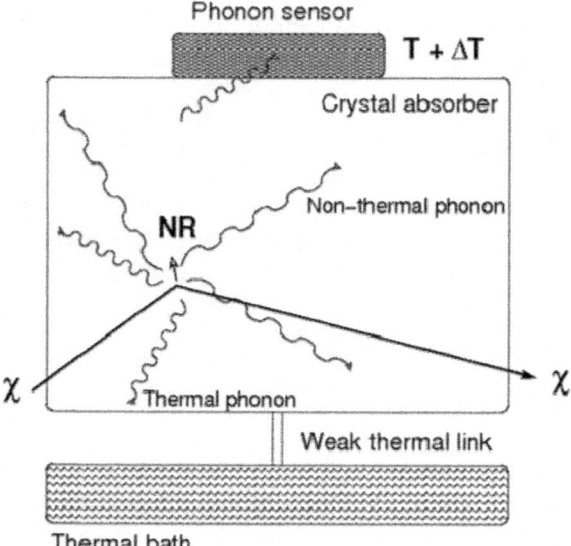

Figure 12: Schematic of a cryogenic phonon detector: an energy deposition E from a nuclear recoil (NR) in an absorber of capacity C(T) produces a temperature rise ΔT which is measured by a thermal sensor.

The crystal is weakly thermally coupled to a heat reservoir which is kept at a constant temperature of about (10–100) mK by a refrigerator cryogenic system. A temperature sensor measures a temperature evolution which can be expressed [283] as

$$\Delta T = \frac{E}{C(T)} \cdot \exp(-t/\tau),$$

(16)

where C(T) is the heat capacity of the absorber material and $\tau = C(T)/G(T)$ with G(T) being the thermal conductance of the link between the crystal and the thermal bath. At cryogenic temperatures, C(T) is very small for some materials due to their T^3 dependence of the heat capacity for a dielectric crystal. This small heat capacity results in a relatively large temperature rise ΔT. For example, germanium cooled down to 20 mK shows temperature rises of typically $1\,\mu K$ for nuclear recoils of a few keV.

To measure the temperature rise ΔT, the most commonly used technologies are neutron-transmutation-doped (NTD) germanium sensors and transition-edge sensors (TESs). To produce NTD sensors, small germanium crystals are exposed to thermal neutrons in order to produce a large amount of doping sites modifying the semiconductor properties of the crystal. The resistance of these thermistors changes strongly with temperature and can be constantly monitored by the voltage drop of the bias current running through them. A TES consists of a thin superconducting film, like tungsten for instance, operated at a temperature inside the phase transition between the conducting and the superconducting states. While this operation is demanding due to requirements on the temperature stability of the cooling system, in general a TES shows in comparison to NTDs an increased sensitivity to measure small temperature changes and is sensitive to athermal phonons.

Cryogenic bolometers exploit, in addition to the phonon signal, either the scintillation or the charge signal to provide particle discrimination. Independent of the read-out technology, the phonon signal is unquenched, i.e. linear with deposited energy, and can be used to determine the recoil energy without dedicated measurements of the quenching factors (see section 6).

The CDMS [284] and its successor the CDMS II experiment [245] use germanium and silicon bolometers to search for dark matter. The experiments are located at the Soudan Underground Laboratory (see section 4) and consist of up to 19 Ge and 11 Si detectors in the final configuration, with a mass of 230 g and 100 g, respectively. These Z-sensitive ionization and phonon-mediated (ZIP) detectors, exploit the phonon as well as the charge signal to allow for particle discrimination. In order to measure the athermal phonons TES are used. Information on the position of the interaction is obtained from the TES pulse arrival-times and the relative signal sizes in multiple sensors. This allows one to select a fiducial volume to reduce the background. The dominant background of these detectors arises from events at the surface of the detectors where a reduced ionization yield is observed which leads to a misidentification of electronic recoils as nuclear recoils. However, phonon pulse-shape discrimination allows one to identify surface events with a misidentification rate of 1 in 10^6 electronic recoils [245]. A combined analysis of all CDMS II detectors yields an upper limit on the WIMP–nucleon SI cross-section of 3.8×10^{-44} cm^2 for a WIMP mass of 70 GeV/c^2. Selecting the four germanium detectors with the best noise conditions and lowest energy thresholds allows one to search for nuclear recoils in the energy range of (3–14) keV$_{nr}$. As a result, a WIMP mass of 7 GeV/c^2 with a cross-section of 10^{-41} cm^2 is excluded [212]. The analysis [193] using silicon crystals is based on an exposure of 23.4 kg d for a nuclear recoil energy range of (7–100) keV$_{nr}$. In this data set, an excess of events above the expected background is observed which corresponds to a WIMP mass of 8.6 GeV/c^2 and a cross-section of 1.9×10^{-41} cm^2. The SD interpretation of the CDMS results can be found in [284]. CDMS II performed additionally a study for an annual modulation of the event rate using data from October 2006 to September 2008 [285]. No evidence for an annual modulation was found and these data disfavour the modulation claim of the CoGeNT experiment [194] which also uses a germanium target.

The successor of the CDMS II experiment is the SuperCDMS detector which employs an improved interleaved ZIP technology (iZIP). These bolometers use an interleaved structure of the phonon and ionization electrodes at the top and bottom faces of the crystals. This allows one to improve the surface event rejection by using the asymmetry of the charge collection [173]. SuperCDMS uses 15 Ge crystals with masses of 0.6 kg each and are sensitive to nuclear recoils between $(1.6-10)$ keV$_{nr}$. A total of 577 kg d science data were recorded focussing on dark matter masses below 30 GeV/c^2, and a limit on the cross-section for an 8 GeV/c^2 WIMP mass of 1.2×10^{-42} cm^2 was derived [203]. Another development of the CDMS collaboration is the CDMSlite (CDMS low ionization threshold experiment) detector which uses a single crystal from the SuperCDMS detector for a dedicated WIMP search with a low energy threshold [286, 287]. In this operation mode, the bias voltage is increased in order to exploit the Neganov–Luke amplification of the phonon signal due to the drift from electron–hole pairs in the crystal lattice. Thus, the energy threshold could be significantly reduced to 56 keV$_{ee}$ and an increase of the energy resolution is observed. However, in this operation mode the simultaneous measurement of the phonon and charge signal is not possible, thereby losing the ability to discriminate between nuclear and electronic recoils. The results of SuperCDMS and CDMSlite set most sensitive exclusion limits at low WIMP masses [203, 287] (see also figure15). SuperCDMS is also the first direct-detection experiment which derives limits on more general WIMP interactions calculated with a non-relativistic EFT (see section 5) [278]. The second generation of the SuperCDMS experiment will be located at SNOLAB having up to a few hundreds of kg target material [288]. This upgrade aims to improve the energy resolution by a factor of 10 and a reduction of the background level by 200. The increase in target mass will be achieved by using 1.4 kg Ge and 615 g Si crystals.

A similar detector concept is used by the EDELWEISS collaboration which operates detectors at LSM. In contrast to the iZIP detectors, in EDELWEISS the signal is measured by thermalized phonons with NTDs. Since thermal phonons do not carry information about the spatial interaction inside the crystal, and surface events dominate the background for a WIMP search, an interleaved structure of the charge read-out is used. The EDELWEISS-II detectors use this technique to identify surface events with a reduced ionization yield, enabling a rejection factor of more than 10^4 [174]. In the final analysis of the EDELWEISS-II configuration, an array of ten 400 g Ge detectors were operated and a total exposure of 384 kg d was achieved [289]. A relatively high energy threshold of

20 keV$_{nr}$ was used in the analysis to avoid a reduced particle discrimination at low recoil energies. The most sensitive limit can be derived at a WIMP mass of 90 GeV/c^2 and a cross-section of around 4×10^{-44} cm^2. A dedicated analysis for low mass WIMPs is performed in [290], using only 4 Ge crystals which show a small background as well as a low energy threshold. The choice of a smaller target results in a lower exposure of 113 kg d; the energy threshold, however, could be reduced to 5 keV$_{nr}$. Thus, limits on dark matter interactions could be derived for masses around 10 GeV/c^2 and a cross-section of 10^{-41} cm^2. The upgraded EDELWEISS-III detector features improved shielding, material selection and background rejection due to an optimized interleaved structure of the electrodes. With increased crystals of 800 g, first results using a single detector with an exposure of 35 kg d exclude 7 GeV/c^2 dark matter particles with a cross-section of 1.6×10^{-41} cm^2 [291].

Both the CDMS and the EDELWEISS experiments have performed axion searches as described in section 7.2 for the CoGeNT experiment. The limits derived for the axion coupling to electrons are summarised in [128, 129] for CDMS and EDELWEISS, respectively.

The CRESST-II experiment at LNGS exploits, in addition to the phonon signal, the scintillation light emitted by recoils in CaWO$_4$ crystals [292]. The phonon as well as the scintillation signal are read out by two optimized tungsten TES. Since particle discrimination solely relies on the scintillation signal, it is necessary to achieve an effective collection of the generated photons. Thus, the housing of the crystals as well as the crystal surfaces are optimized to avoid an absorption of the photons or inner total reflections. First limits on dark matter interactions have been derived already in 2004 [292]. The second phase of CRESST-II had, in addition to a larger array structure for crystals, improvements of the neutron shield and an active muon veto. A total exposure of 730 kg d with eight detectors was achieved, where each crystal weighs about 300 g and shows an energy threshold in the range from 10.2 keV$_{nr}$ to 19.0 keV$_{nr}$. An excess of events is observed, corresponding to a WIMP mass of 11.6 GeV/c^2 (4.2 σ) or 25.3 GeV/c^2 (4.7 σ) with a cross-section of 3.7×10^{-41} cm^2 or 1.6×10^{-42} cm^2, respectively [192]. It is worth mentioning that the main background in this analysis is due to collisions of lead nuclei with the crystal from ^{210}Po α-decays where the emitted α remains undetected. A further improvement of the detector layout increased the efficiency to measure the emitted α events from ^{210}Po decays, leading to a strong suppression of the main background. In addition, the detectors show an improved phonon and photon read-out efficiency, leading to a significant reduction of the energy threshold to 600 eV$_{nr}$. With 29.35 kg d of exposure the previous signal claim could not be verified. A sensitivity to WIMP masses below 3 GeV/c^2 was reached, while no background event in the signal region was observed [175]. With the same detector technology and exposure a dedicated low mass analysis was performed excluding WIMP interactions for a mass of 3 GeV/c² at a cross-section of 8 × 10⁻⁴⁰ cm² [204]. Using the detector module with the lowest energy threshold of 307 eV and an increased exposure of 52 kg d, a cryogenic bolometer showed for the first time sensitivity to sub-GeV/c² dark matter masses at the cross-section level of 10⁻³⁷ cm² [293]. In the future, the CRESST collaboration will focus on low mass WIMP detection by reducing the crystal size (24 g) and lowering the energy threshold [294] to less than 100 eV. This would allow one to search for WIMP masses down to 1 GeV/c². Another possible detector improvement is also considered to make use of the Neganov–Luke amplification of the phonon signal to lower the energy threshold [295]. Also the ROSEBUD experiment uses scintillating bolometers to search for dark matter [296]. First results using sapphire crystals in the Canfranc Underground Laboratory showed promising results but due to the high background and the small exposure the results were not competitive.

A possible next-generation experiment, the European Underground Rare Event Calorimeter Array (EURECA), aims to build a facility to operate 1000 kg of cryogenic detectors, both $CaWO_4$ and Ge detectors [297]. This experiment is a joint effort mostly originating from the EDELWEISS, CRESST and ROSEBUD collaborations. The detectors would be located at the LSM laboratory, consisting of 150 kg target material in a first phase, followed by a second phase with 850 kg. The final goal is to reach a sensitivity of 3 × 10⁻⁴⁶ cm². In principle, a joint experiment between EURECA and SuperCDMS would be feasible, combining the various mentioned technologies and exploiting their complementarity.

Liquid Noble-Gas Detectors

Liquid noble-gas detectors offer the advantage of large and homogeneous targets with high scintillation and ionization yields. Currently, LAr and LXe detectors are used as detector media. There are also some R&D activities carried out for liquid neon [298]. The scintillation of both LAr and LXe is in the ultraviolet regime at 128 nm and 175 nm, respectively [299]. While for LAr it is common to use wavelength shifters and detect light in the blue wavelength region (~400 nm), in LXe the photons can be detected directly by using photosensors with windows made out of quartz which is transparent to the xenon scintillation light. After the passage of ionizing radiation, ionization or excitation of the medium takes place. The excited or ionized atoms form excimers, D_2^* or D_2^+ which de-excite, emitting ultraviolet photons. The free electrons which appear in the ionization can either recombine to produce further scintillation light or can be extracted with a drift field to be collected as an additional signal [300]. Furthermore, LXe has the advantage of containing almost 50% of non-zero spin isotopes, ^{129}Xe and ^{131}Xe, providing additional sensitivity to SD WIMP interactions [210]. The high density of xenon (about 3 gℓ^{-1}) provides excellent self-shielding such that a radio-clean innermost volume can be selected for analysis.

In order to distinguish the main background due to γ and e^- interactions (electronic recoils, ER) from the interactions of WIMPs with nuclei (nuclear recoils, NR), two methods can be applied in liquid noble-gas detectors: pulse-shape discrimination and charge-to-light signal ratio. The short- (singlet) and long-lived (triplet) states that produce the luminescence in these media are populated at different levels for different types of particles. This results in a differentiation between ER and NR. This technique gives large separation power in liquid argon due to the easily separable lifetimes (6 ns and 1.6 μs) [301] of the two components. However, pulse shape discrimination provides a good separation only for a large number of measured photons and therefore, a higher energy threshold has to be considered. In LXe, the values for the decay constants are too close to each other, 4ns and about 22 ns [302], giving less rejection power.

Single-phase (liquid) detectors consist typically of a spherical target, containing the liquid medium, which is surrounded by photo-detectors (see figure 13 left). A main advantage is the 4π-photosensor coverage which results in a larger light output compared to detectors which are only partially instrumented. The distribution and timing of the photons at the photosensors can be used to determine the position of the event typically with ~ cm resolution enabling the definition of a fiducial volume. Pulse shape is the main particle-discrimination parameter in single-phase detectors. DEAP [303] and CLEAN [304] are examples of experiments using liquid argon in single phase. They are currently being commissioned at the SNO laboratory in Canada. Both detectors use light guides from the medium to the photosensor sphere in order to minimize the impact of the background from the radioactivity of the sensors. The DEAP-I prototype showed a discrimination power of 10^{-8} and an acceptance of 50% for nuclear recoils above an energy of 25 keV [305]. The background of this experiment can be mainly explained by radon daughters decaying on the surface of the active volume, misidentified electronic recoils due to inefficiencies in the pulse shape discrimination, and leakage of events from outside the fiducial volume [306]. These backgrounds will be strongly reduced in the DEAP-3600 detector due to its higher light yield and simpler geometry compared to the first prototype. At the time of writing, DEAP-3600 is being commissioned. The pulse-shape discrimination mechanism also works for LXe [307] but not as efficiently as in the argon case due to the similar decay components of the short- and long-lived states. The XMASS experiment [308] in Japan employs the single-phase technology with about 800 kg of LXe. Ultra-low radioactive materials are used for construction to further reduce the experimental background. In the data acquired during 2012, an unexpected radioactive contamination originating from the photosensors appeared. Nevertheless, some results on low WIMP masses [309] and on inelastic scattering off xenon [120] have been derived. The detector has meanwhile been refurbished to shield the contamination from the PMTs and new data are expected within 2015. In a next phase, XMASS1.5 plans to extend to 5 ton LXe mass with about 1 ton fiducial mass [310]. On larger time-scales, XMASS2 is proposed as a multi-ton ($\sim$24 ton) multi-purpose detector [311].

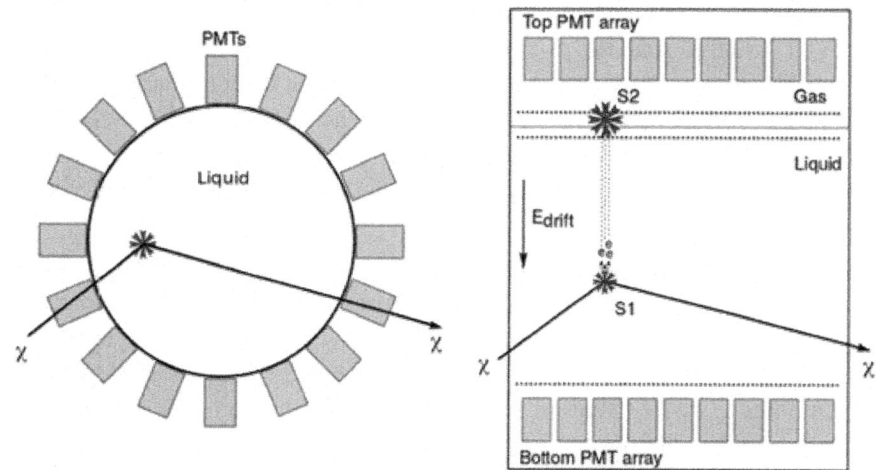

Figure 13: Schematic of single-phase (left) and double-phase (right) liquid noble-gas detectors.

A second method, the double-phase detectors (liquid and gas), enables one to detect both the scintillation light and the charge signal from ionization produced by an energy deposition [300]. The ratio of the two signals depends on the particle type and allows one to separate signal-like events from background ones. Typically, two arrays of photosensors, on the top and bottom of the detector, are employed to detect the prompt light signal. Ionized electrons drift upwards to the liquid–gas surface and are amplified via proportional scintillation in the gas phase [312] which is also measured by the photosensor arrays. Therefore, double phase detectors are operated as a TPC (see figure 13 right). Position reconstruction of events is performed obtaining the z component from the time difference between the scintillation signal and the charge signal and by using the light pattern in the photosensors for the (x, y)-coordinates. The typical position resolution is in the order of millimetres. WARP [247], operated during 2005–2006, was the first LAr detector which produced dark matter search results. It was located at the LNGS laboratory in Italy and consisted of 2.3 ℓ liquid argon. Currently, the DarkSide experiment [313] is operating with about 50 kg active mass and first results have shown a large light yield at ~8 photoelectrons (PE) per keV energy which results in a very good separation of signal from background ($>1.7 \times 10^7$) using the information contained in the pulse shape. The energy threshold for this analysis was placed at 38 keV$_{nr}$. Given the null results of this first run, an exclusion limit is placed which is at 6.1×10^{-44} cm^2 at 100 GeV/c^{2}WIMP mass. On the long-term DarkSide plans on a multi-ton detector featuring 3.6 tons in the target volume [314]. A ton scale detector, the ArDM [315, 316], was first tested at CERN and it is being commissioned in 2015 at the Canfranc underground laboratory in Spain.

The LXe TPCs used by the ZEPLIN [317] and XENON10 [318] experiments showed from 2006 to 2011 the potential of this technology to search for dark matter. The exclusion limits on the coupling of dark matter particles to nuclei placed by these detectors were most constraining at that time [200, 319]. The ZEPLIN detector, which operated at the Boulby underground laboratory, achieved a high separation between signal and background events by using a flat detector geometry allowing one to increase the electric field in the liquid to a maximum of almost 9 kV cm^{-1} [317]. The γ-ray rejection factor was at 5×10^3 for the energy range (2–16) keV$_{ee}$ [246]. Besides the common SI and dependent results, the XENON10 experiment performed a study using the charge signal (S2) alone. This allows one to lower the detection threshold down to ~1 keV$_{nr}$ but gives up the possibility to discriminate signal and background. In this mode, the LXe technology obtained competitive sensitivities at WIMP masses as low as 5 GeV/c^2 [320] (see figure 15). The successor of XENON10, XENON100, started operation at the LNGS laboratory in 2009. Its total mass of LXe is 161 kg, where 62 kg are contained inside the TPC and the rest is used for a LXe veto surrounding the TPC. The longest dataset for which results have been released consists of 225 live days [321]. Self-shielding is employed to minimize the background and, therefore, a 34 kg as fiducial volume for the analysis. No evidence for dark matter was found. When interpreting the data as SI interactions of WIMP particles, a best sensitivity of 2×10^{-45} cm^2 for 55 GeV/c^2 mass is derived. Natural xenon contains two nonzero nuclear-spin isotopes, ^{129}Xe and ^{131}Xe, with an abundance of 26.4% and 21.2%, respectively. Therefore, the absence of events above the background prediction also allows one to exclude WIMP interactions which depend on the nuclear spin [210]. Both SI and dependent exclusion limits are shown in section 8. Finally, the electron-recoil part of the data has been investigated in order to search for axion-induced signatures. Most sensitive upper limits on the coupling of

axions to electrons are derived ($g_{Ae} < 10^{-12}$ at 90% C.L.) between 5 and
10keV/c² axion masses [130]. In addition, the XENON100 electron-recoil data
have been used to study possible periodic variations of the event rate, allowing
one to exclude the DAMA annual modulation at $4.8\,\sigma$ [322]. Furthermore,
exploiting the low background rate of the experiment, various leptophilic dark
matter models have been excluded as explanations of the DAMA signal [323].
To further increase the sensitivity, a next generation detector, XENON1T
[324], consisting of about 3 tons of LXe, is being constructed. The goal is
reach two orders of magnitude improvement in sensitivity by also reducing
the background by a factor of ~100 compared to XENON100. XENON1T is
built such that the main part of the infrastructure can host ~6 tons of LXe and
therefore, an upgrade to XENONnT can be performed with a moderate effort.

In 2013, the LUX experiment [159], installed at the Sanford underground
laboratory in the US, released first data [211] of 85 live days. These results
were derived using 118 kg target mass and improved the results of XENON100
down to 7.6×10^{-46} cm² for a WIMP mass of 33 GeV/c². This is currently
the lowest limit for direct-detection experiments for SI interactions for WIMP
masses above 6 GeV/c². In particular, these results are also in strong tension
with signal indications from other experiments. The experiment has achieved
a very low electronic recoil background at the level of 3.6×10^{-3} events/(keV
kg d) [181]. Due to the larger light yield of the detector (at 8 PE/keV at 662
keV energy), 2.5 times higher than in XENON100, the experiment has set
strong constraints at low WIMP masses. More results from LUX are expected
within 2015. The LUX and ZEPLIN collaborations have joined to build the
multi-ton LZ detector hosting about 7 tons of LXe in the target volume [325,
326] increasing, thereby, the sensitivity on WIMP-matter cross-sections. The
LXe TPC technology is also used in the Chinese PandaX [327] experiment
which is operated at the Jin-Ping underground laboratory. In the first phase of
the experiment, the target volume consists of 120 kg. The most recent results
from 2015 used 80.1 live days exposure with a fiducial volume of 54 kg. The
experiment tested cross-sections down to 10^{-44} cm² for a 45 GeV/c² WIMP
mass [328]. Due to the pancake-like geometry, the light yield is rather high
(1.5× larger than XENON100), but also the background is increased. The
background rate is more than six times larger than that of LUX because in
the flat geometry the background from the PMT arrays is closer to the fiducial
volume. PandaX plans to increase the height of the TPC in order to measure
with a mass of 500 kg. In a final step, the detector will be upgraded to host a
multi-ton target [327].

Superheated Fluids

Bubble chambers were often used in past decades in accelerator experiments until new technologies such as, for instance, gaseous detectors provided a better performance. Over recent years, the technology of using superheated liquids has been revived in the context of dark matter searches [329]. This branch of experiments can be divided into bubble chambers and droplet detectors. Both technologies use refrigerant targets operated in a superheated state slightly below its boiling point. Interactions of particles with the target can be observed by the induced process of bubble nucleation. To create an observable bubble in the detector a phase transition of the medium is necessary. Therefore, the deposited energy by the particle must create a critically sized bubble, requiring a minimum energy deposition per unit volume. This process can be described by the 'hot spike model' [242]. An event is then photographed with charge-coupled device (CCD) cameras, and the position of the bubble can be determined with ~mm resolution. This allows us to define an innermost volume for the analysis, featuring lower background. After the formation of each bubble, the medium has to be reset by a compression of the liquid phase followed by a decompression to a value below the vapour pressure. In contrast to bubble chambers, droplet detectors make use of a water-based cross linked polymer to trap the bubbles resulting in a shorter dead time of the detector [330].

The major advantage of this technology is that, being close below the temperature of the phase transition, bubble chambers are insensitive to minimum ionizing backgrounds which generally dominate the backgrounds of other dark matter detectors. In this way, most of the backgrounds created by γ-rays, x-rays and electrons from β-decaying isotopes are avoided. The remaining radiation which is able to produce nucleation are α-particles, nuclear recoils from neutron interactions and WIMP-induced recoils. Due to the explosive character of the phase transition, acoustic signals can be used to discriminate α-background events. For instance, the COUPP experiment has shown a <99.3% efficiency in rejecting α-events, as they produce louder acoustic emission than nuclear recoils [243]. Similarly, the rise-time and the frequency of the acoustic signal is used in the PICASSO experiment to mitigate the α-background [331]. Although bubble chambers are threshold devices, i.e. counting events above a certain energy, by varying the temperature and/or pressure, the energy threshold can be changed. Existing detectors achieved energy thresholds of the order of a few keV nuclear recoil energy. Typically, the targets being used (CF_3I, C_2ClF_5, C_3ClF_8 and C_4F_{10}) contain fluorine which has an unpaired number of protons and is thus sensitive to SD interactions.

Moreover, fluorine has a particularly large expectation value for the proton spin content which enhances the sensitivity for SD interactions to protons [108].

Four different experiments have been operating during the last years using the bubble chamber (COUPP [243] and PICO [205]) and droplet detector (PICASSO [331] and SIMPLE [201]) technologies. The used target masses reached only a few kg, hence they are not competitive in the SI interpretation of the data. Nevertheless since their target contains fluorine, these detectors are sensitive to proton-coupling SD interactions. Experiments using germanium or LXe have unpaired neutrons and consequently lower sensitivity to proton-coupling. As a result, the results of bubble chamber detectors have best sensitivities within this interpretation of the data (see figure 16, right). One of the first bounds on the dark matter cross-section from a detector using superheated fluids was achieved by the SIMPLE experiment which is operated at LSBB in France (see section 4). It used 215 g of C_2ClF_5 as a target and reached exposures up to 13.7 kg d. With an energy threshold of 9 keV, a sensitivity to the SD WIMP–proton cross-section of 5.7×10^{-39} cm^2 at 35 GeV/c^2 was achieved [201]. Among the above-mentioned detectors, PICO (formed from the PICASSO and COUPP experiments) shows the strongest exclusion limit on the SD WIMP–proton cross-section at 40 GeV/c^2 of around 9×10^{-40} cm^2 at 90% C.L. PICO is operated in the SNOLAB underground laboratory (see section 4) with a mass of 2.9 kg of C_3F_8 reaching a total exposure of 211.5 kg d and an energy threshold ranging from 3.2 to 8.1 keV. It should be mentioned that the PICO detector is able, for the first time, to discriminate efficiently α-events by the acoustic signal due to the efficient bubble nucleation processes in C_3F_8 [205]. For SI interactions, PICO is competitive with other detector technologies at low WIMP masses (<6 GeV/c^2) due to the low energy threshold and light target nuclei. For higher WIMP masses, however, the small exposure limits its sensitivity.

Directional Detectors

In the previous sections, dark matter signatures based on either the annual modulation of the recoil rate or on the spectral shape of the signal are exploited. An additional possibility, introduced in section 3.1, is to measure directly the recoil track produced by the dark matter interaction. This would provide information on the ionization density (dE/dx) as function of position, on the range and eventually on the direction of the recoiling nuclei. In the reference frame of the Earth, the WIMPs of the Milky Way halo are expected to originate from a preferred direction, approximately the Cygnus constellation. Therefore, an asymmetry in the number of events scattering forwards and backwards is expected [101].

The range of dark matter-induced nuclear recoils is below 100 nm for energies < 200 keV in liquids and solids, making the track reconstruction very challenging. The most promising strategy for directional searches is, instead, the use of low pressure gases such that the track length of the induced recoiling nucleus is large enough to be resolved. For a pressure < 100 Torr (< 130 mbar), the range of a WIMP-induced recoil with a mass of 100 GeV/c^2 and a speed of 220 km s^{-1} is, for instance, (1–2) mm. Note that this range varies significantly with the gas pressure. Such low pressures result in a low target mass and, consequently, very large detectors are necessary to achieve sensitivities comparable to the experiments mentioned before. However, the measurement of the nuclear recoil direction would constitute the ultimate confirmation of dark matter detection.

Current directional developments use the gaseous TPC technology for directional dark matter searches [332]. The drift gas serves as target material and detector simultaneously. Commonly used gases are CS_2, CF_4 and ^{3}He, where the last two are favoured due to the unpaired nucleons that give sensitivity to SD interactions. The ionization charge produced after a nuclear recoil is drifted by a homogeneous field to the read-out plane. Using the ionization pattern, the (x, y)-projection of the recoil can be reconstructed. The extraction of the z-projection is dependent on the detector and is discussed for each detector below. Angular resolution, i.e. the precision of the reconstructed angle, is around 30°. Figure 14 shows a schematic view of a gaseous TPC. Several read-out technologies are being developed including multi-wire proportional chambers (MWPCs), CCD cameras or gas electron multipliers (GEMs [333]). The amplification of the drifted charges before the read-out plane allows one to lower the detection energy threshold of the experiments.

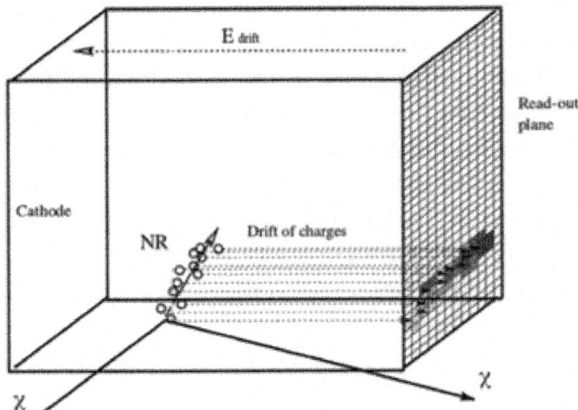

Figure 14: Schematic of a track reconstruction in a directional low-pressure gaseous time-projection chamber (TPC).

In general, all charged particles produce tracks in the gaseous target. However, gamma- or beta-induced electronic recoils can be distinguished from nuclear recoils by determining the track length. The length for electronic recoils is at all energies about a factor of 10 larger than for the equivalent nuclear-recoil energy. In contrast, alpha-induced tracks are not as well separable from nuclear recoil tracks [61]. Therefore, a low alpha contamination is necessary. The energy threshold of these detectors is coupled to the energy to create an electron–ion pair (called W-value). These W-values are in the range of tens of eV which should, in principle, enable to reach sub-keV energy thresholds.

The DRIFT-II experiment [334], currently the largest directional detector, is operated at the Boulby underground laboratory in the UK (see also section 4.2). The detector is a TPC read-out by a MWPC. Its volume of 0.8 m^3 is filled with a low-pressure mixture 30:10:1 Torr of $CS_2{:}CF_4{:}O_2$ gas. This mixture provides 33.2 g of fluorine in the active volume as target mass for SD WIMP interactions. The DRIFT detector uses the drift of negative ions (in particular CS_2^-) which travel at different velocities to determine the z-component (drift direction) of the event. This enables one to fiducialize the volume by choosing the cleanest region for the analysis. Another advantage of using CS $_2$ is the reduction of diffusion while drifting. Using this technique, the DRIFT collaboration produced at the end of 2014 a background-free directional result for SD (proton coupling) interactions. This result is the most sensitive for directional searches but it is not yet competitive with the leading results from bubble chambers [205]. The MIMAC experiment [335] also aims to measure the nuclear recoil energy and its angular distribution. Since 2012, a prototype-chamber has been operated at the Modane underground laboratory (LSM) in France with 50 mbar of a mixture with CF_4, 28% CHF_3 and 2% C_4H_{10}. The read-out of the ionization electrons consists of pixelated micromegas [336]. The initial acquired data showed that a main background is due to the decays of radon. These data have been used to show, for the first time, the observation of a low energy nuclear recoil originating from the α-decay of radon [337]. In addition, a parameter based on the diffusion size of the electron cloud is defined to allow for a z-component determination which in turn allows one to define a fiducial volume. The DMTPC experiment [338] is a planned m^3-scale TPC using CF_4 at 50 Torr. The detector is a TPC with a charge amplification region. The primary ionization drifts to this region where an avalanche with a gain of 50 000 takes place to amplify the signal. Scintillation photons from ion-recombination in the amplification region are acquired with CCD cameras. The image allows one to determine the track geometry and the direction of the recoil. First prototypes are tested to measure both the energy and the direction of nuclear recoils [339]. Finally, there are two further directional gas-TPCs where the readout is based on the GEM technology: the NEWAGE [340] and the

D^3 [341] experiments. While the former, located at the Kamioka underground laboratory in Japan, has performed two science runs in 2010 and 2013 [342], the latter is in the prototyping phase [343]. The goal of the experiments mentioned above is to achieve volumes of $\sim m^3$ with the required signal topology. The increase of mass would then be realised by a multi-module detector.

Beside the low-pressure gaseous detectors mentioned above, nuclear emulsions can be used to reconstruct sub-micrometer particle tracks. Fine grained emulsions using silver-halide crystals of several tens of nm have been produced [344] and their tracking capabilities have been shown using nuclear recoils from a neutron source. The read-out is performed via optical and x-ray microscopes and an angular resolution of about 20° has been achieved. The read-out efficiency for 120 nm long tracks is larger than 80%. After an initial R&D phase, plans to operate a detector at LNGS are on-going.

Novel Detectors

R&D activities are on-going in the development of detectors made out of solid xenon [345]. Besides some of the advantages of xenon mentioned in section 7.4, solid xenon shows an increased amount of light collection and a faster electron drift compared to the detectors using liquid phase. In addition, if it were possible to read a phonon signal, the energy resolution and the energy threshold were greatly improved. Moreover, it would open the possibility to measure all three signals (ionization, scintillation and heat) which eventually results in an increased particle discrimination power. It has been noted, however, that the development of crystals at sub-Kelvin temperatures might be challenging. So far, the scalability of transparent solid-xenon detectors to masses in the kg-scale has been demonstrated by a 2 kg crystal which was grown at a temperature of 157 K.

DAMIC [346] is a detector using silicon CCDs to search for light WIMPs with (1–10) GeV/c^2 masses. Due to their low electronic noise, these devices can be operated with thresholds as low as 40 eV_{ee}. First studies of the radioactive contamination of these devices show that the levels are sufficiently low to reach a competitive sensitivity at low WIMP masses (see first results in figure 15). Currently, a low-radioactivity 100 g detector is being installed at SNOLAB [347].

An improvement of detectors using superheated fluids as the detection medium is the 'geyser technique' (or condensation chamber) exploited by the MOSCAB experiment [350]. While showing the advantages of the droplet and bubble chamber technologies (see section 7.5), the geyser technique allows one to reset the detector within a few seconds.

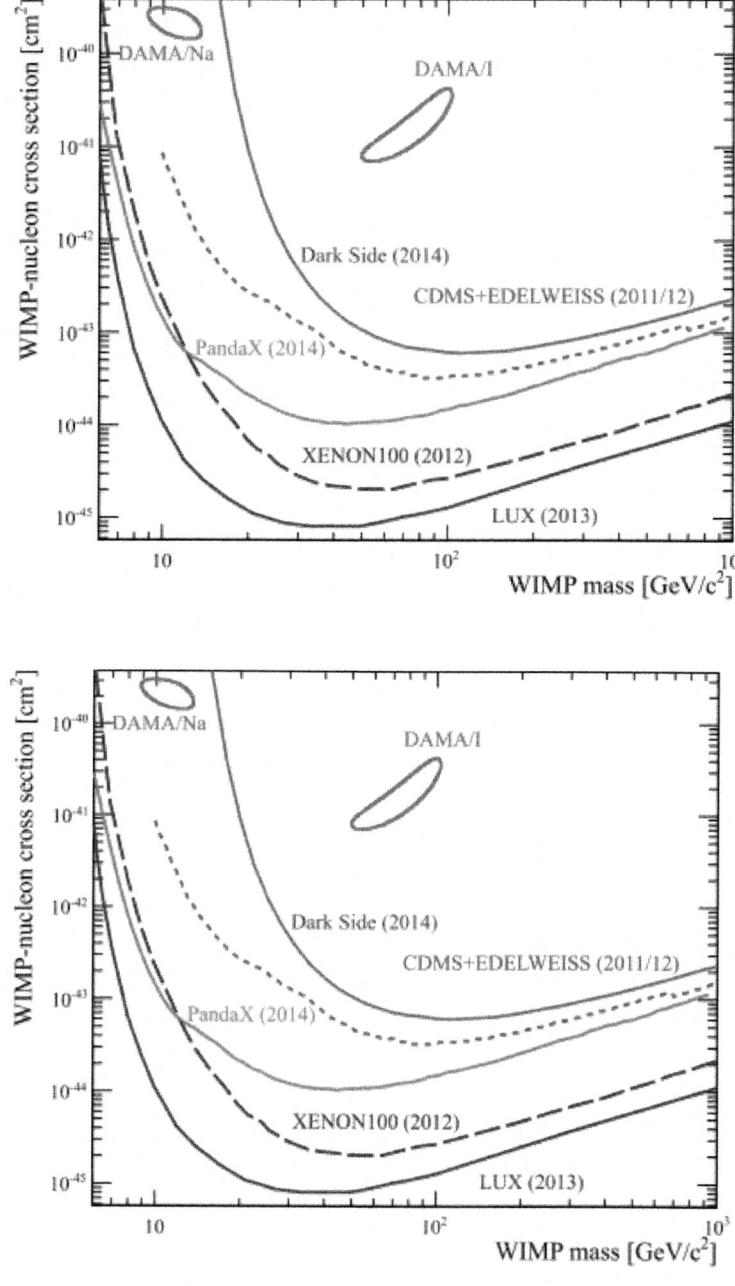

Figure 15: Overview of signal indications and exclusion limits from various experiments for spin-independent WIMP–nucleon cross-section for low WIMP masses (left) and high WIMP masses (left). Data from dmtools [352] or private communications.

This technology not only simplifies the overall detector set up significantly, allowing one to increase the target mass, but also reduces the dead time of the detector. To date, a prototype detector of 0.5 kg (C_3F_8) is operating and a 40 kg detector is being developed. Already the 40 kg detector is expected to surpass the current best sensitivity by the PICO experiment by several orders of magnitude. For the future, extensions are planned up to 400 kg allowing one to probe SD proton interactions down to 10^{-43} cm^2.

An interesting proposal is to use detectors made out of DNA or RNA to search for dark matter [351]. A possible realization would consist of thin gold films with strings of nucleic acids hanging from it. A gold nuclear-recoil produced by the interaction of a WIMP, would create a break in the sequence of these strings. The location and geometry of the break can be reconstructed by techniques common for biologists. By using the track reconstruction, the directionality of the signal can be employed to reduce the background. A threshold at ~0.5 keV would allow one to focus on the low-WIMP mass region. The aimed target mass would consist of about 1 kg of gold. Note that this technology has, so far, not been used in any astroparticle physics-related experiment and the feasibility of such a detector has to be experimentally shown.

SUMMARY AND PROSPECTS FOR THE NEXT DECADE

In order to prove the existence of weakly interaction massive particles (WIMPs), experimental efforts can be categorized into indirect detection, i.e. via secondary particles created by dark matter self-annihilation, production of dark matter in particle colliders and direct detection of dark matter scattering off a target. This review summarises the main concepts of direct detection experiments, namely dark matter detection signatures, methods for background reduction, detector calibrations, the statistical treatment of data and the interpretation of results. The focus lies on section 7 where various technologies aiming to directly detect dark matter interactions are discussed together with their current status and plans. In the following, some of the possible interpretations of results are presented and prospects for the near future are discussed.

WIMP interactions with the target of an experiment can be detected by a characteristic energy spectrum, an annual modulation of the measured event rate or by a directional dependence of interaction tracks (see section 3). Figure 15 compiles signal indications and exclusion limits for both low WIMP masses (left) and high WIMP masses (right). Signal indications stated by several experiments are shown as closed contours, whereas limits are represented by curves excluding the parameter space above.

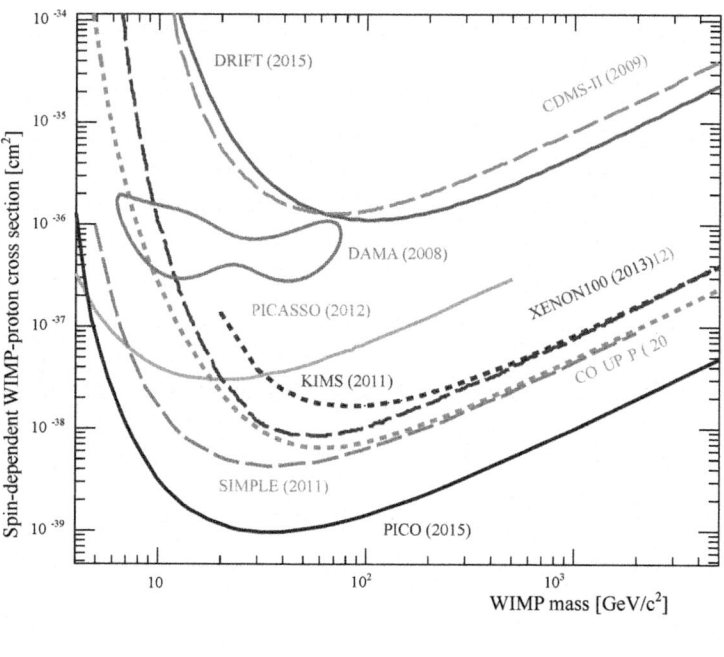

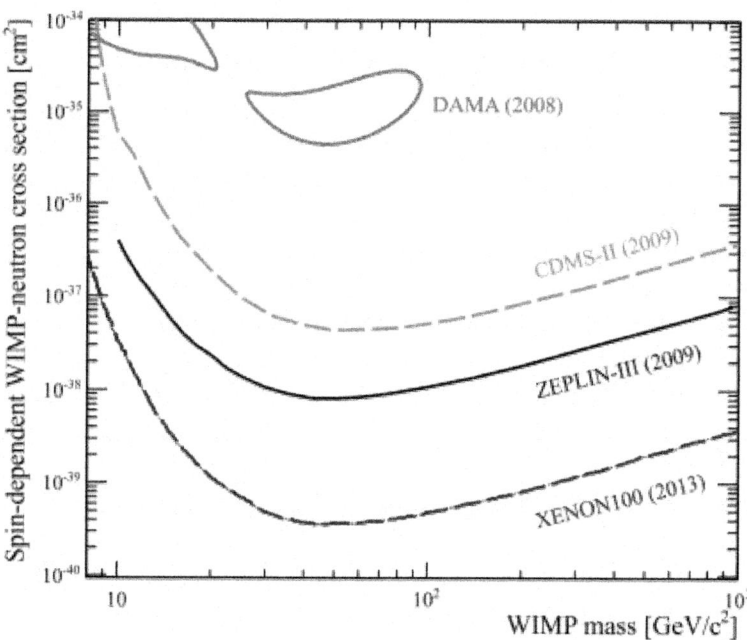

Figure 16: Exclusion upper-limits for spin-dependent WIMP-nucleon cross-section assuming pure proton coupling (left) and pure neutron coupling (right). Data from dmtools [352] or private communications.

The separation of the tested parameter space in the two WIMP-mass ranges have become more important in recent years, since various experiments have started to focus on a particular mass scale to exploit the specific advantages of the individual technology (see section 7). For experiments showing sensitivity to low WIMP masses, the determination of the energy threshold becomes a crucial aspect. For this purpose, dedicated measurements of the target energy scale are performed. The systematic uncertainties in the determination of these scales can affect, indeed, the results shown in figures 15 and 16. In section 6, the calibration strategies for various detector types are summarised.

Only a few experiments analysed the data for an annual modulation of the event rate, mainly due to the requirement to achieve a long-term stability of the detector. The annual modulation of the rate measured by the DAMA experiment has a significance of 9.3 σ [251] and, depending on the analysed target atom (Na or I), the derived signal regions (solid red) are shown in figure 15. The origin of the signal, however, remains controversial, especially since results of other experiments are in strong tension with the DAMA claim. In section 7.1, possible explanations for the DAMA signal are discussed including both dark matter and non-dark-matter related origins. Using a germanium detector as the target, an annual modulation of the signal at the level of 2.2 σ (solid dark green) was also claimed by the CoGeNT collaboration [270]. A reanalysis of the data with different background assumptions shows, however, even lower significances [271, 272]. In addition, the data from the CDMS II detector could not verify a modulation of the measured event rate by evaluating the germanium data [285].

Commonly, the spectral shape of signal candidate-events is used to constrain dark matter interactions with the assumption of SI (and isospin-conserving) elastic scattering off WIMPs. In 2013, the presence of three observed events in the CDMS silicon detectors was above the expected background [193] (solid orange). Although no further data using silicon have been released so far, the SuperCDMS collaboration performed a science run in 2014 with improved low-threshold germanium detectors [203] which cannot verify the previous signal (dashed brown). Shortly before, an event excess measured by the CRESST experiment in 2012 [192] could be interpreted by WIMP interactions with masses of 11.6 GeV/c² (4.2 σ) or 25.3 GeV/c²(4.7 σ). However, new results derived by the same collaboration using an upgraded detector with improved background conditions and the same target element could not reproduce this excess [175] (solid green). Therefore, the initial signal claim is not shown in figure 15. In addition, the results from the LXe detectors XENON100 [321] (dashed blue) and LUX [211] (solid violet) disfavour the signal indications described above. The presence of various dark matter indications in the region around ~10 GeV/c^2 created some excitement; however, meanwhile improved results from several experiments indicate that probably in most cases, background was responsible for the observed events. This emphasizes the relevance of the background prediction and the quantification of its uncertainty. The presented experimental results are, in addition, derived with different statistical frameworks which consider systematic and statistical uncertainties to different degrees. Section 5.2 discusses briefly the implicit assumptions made in the various statistical frameworks used by the experiments. In conclusion, these figures show the strength of detector technologies with a low energy threshold (e.g. cryogenic bolometers, CCD) since they are most constraining for WIMP interactions with masses below 5 GeV/c^2. In contrast, LXe TPCs have the highest sensitivity for larger dark matter masses due to their large target masses. The third signal indication of dark matter interactions is given by a directional dependence of the interaction tracks (see section3). Low pressure gaseous detectors aim to measure the direction of the recoil atoms (see section7); however, their exposures are currently not competitive with the sensitivities of other technologies.

As discussed in section 3.2, the results of a dark matter experiment can be also interpreted by SD interactions if the isotopes in the target material contain an unpaired number of nucleons resulting in a non-zero spin expectation value. It is common to derive results separately for spin couplings to neutrons and protons. Figure 16 shows the SD results from various experiments for pure neutron-coupling (left) and pure proton-coupling (right). To date, XENON100 [210] (dashed blue) shows the strongest limit for SD interactions on neutrons. This is not only due to the larger exposure of the experiment but also because ^{129}Xe and ^{131}Xe have a high neutron spin expectation-value. In contrast, for SD WIMP interactions with protons, experiments using ^{19}F have the highest sensitivities also because of the large spin expectation-value for this isotope. Currently, the most constraining limits are derived from technologies using superheated liquids containing ^{19}F as in the PICO [205] (solid dark blue), COUPP [243] (dotted green) and SIMPLE [201] (dashed green) (see figure 16) detectors, despite their lower exposures. Measuring the directionality of the recoil tracks with low-pressure gaseous detectors containing ^{19}F enables one to search also for SD interactions. The DRIFT experiment sets one of the first competitive limits on SD proton interactions [334] (solid magenta).

Note that the choice of present experimental results interpreted by SD and SI interactions with matter is given by their relative strength in comparison to general coupling terms but are not the only possibilities. A more generalized interpretation of dark matter interactions containing, for instance, also velocity suppressed operators in the context of a non-relativistic EFT is summarised in section 3.2. Although this general approach is not yet widely used, in 2015 first experimental results have been displayed in this framework [112].

Systematic uncertainties in astrophysical parameters of the dark matter halo distribution are entirely neglected in figures 15 and 16. Even though the results are usually derived by a common choice of astrophysical parameters using the standard halo model (see section 3.4), a comparison of different detector targets is demanding. This is caused by the varying kinematics of WIMP interactions on different target elements and the various energy thresholds. A method to display results independently of astrophysical assumptions to avoid a possible bias in the comparison among results is presented in section 5.3. It has to be noted that, even using this method, the discrepancy between the DAMA signal and the null results from other experiments remains.

Although this review focuses on direct detection experiments, a comparison to collider searches and indirect detection experiments is beneficial due to the complementarity of the approaches. This complementarity will be especially relevant in the presence of a signal. Collider searches exploit mostly 'mono-signatures' (e.g. mono-jets, mono-photons) accompanied by missing transverse-energy to constrain dark matter masses. Since no indication of dark matter particle-production at colliders has appeared so far, these results can, in principle, be compared to those of direct-detection experiments. However, a direct comparison between both is not possible and first dark matter limits have to be mapped to a common parameter space. Instead of comparing individual dark matter models, collider searches can be interpreted in an EFT approach (see e.g. [353–355]) or by considering minimal simplified dark matter models (e.g. [356, 357]). It has to be remarked that an EFT approach can only be used for certain dark matter masses and coupling strengths as pointed out in e.g. [358,359]. In general, due to the limited centre-of-mass energy in colliders, direct-detection experiments show a higher sensitivity at heavy WIMP masses. Collider searches are, in turn, most constraining below the energy threshold of dark matter experiments and, hence, at low WIMP masses. Moreover, for SD interactions, direct-detection signatures lose the A^2 enhancement of the event rate (see section 3.2) and colliders, i.e. results from the LHC, have a higher sensitivity at all WIMP masses (see for example [66–68]). Note, however, that collider searches can neither directly measure the dark matter particle nor test its lifetime. Therefore, the definitive confirmation of such detection would only occur in combination with direct-detection results.

Also for indirect detection, a comparison with direct searches is, in general, demanding since the former approach is only sensitive to the thermally averaged self-annihilation cross-section of dark matter. Hence, these processes do not allow us to constrain elastic scattering of dark matter particles to baryons.

However, it is possible to gravitationally capture dark matter particles inside the Sun via elastic scattering. The strength of the elastic scattering cross-section determines the dark matter density inside the Sun, which is in turn proportional to the dark matter pair-annihilation rate. From all possible dark matter self-annihilation products, the only detectable particles that would reach the Earth are neutrinos [360, 361]. Therefore, SI interactions can be constrained by the elastic scattering on solar hydrogen and helium, whereas SD interactions can only be probed by scattering off hydrogen (protons). A few experiments as Super-Kamiokande [93], Ice-Cube [92] and Baksan [362] have exploited these channels to constrain the dark matter cross-section with matter by searching for high energy neutrinos from the Sun. Similar to the collider searches, the constraints from these experiments are not competitive with direct detection experiments for SI interactions, except at very low WIMP masses. SD proton scattering can be, instead, very well constrained, exceeding the sensitivities from direct detection experiments but not the limits from LHC.

Over recent decades, although no definitive evidence for dark matter has appeared, great progress has been achieved in direct dark matter searches. Figure 17 summarises the time evolution of the SI cross-section sensitivity since the first results of a germanium detector in 1985 [269]. The top panel of figure 17 shows upper limits for a 50 GeV/c^2 WIMP mass and the bottom panel for a 5 GeV/c^2 dark matter particle. The data are separated in low and high WIMP-mass regions due to the recent development towards an increase of sensitivity at low WIMP masses. After the first results at high WIMP masses from germanium detectors (black circles), cryogenic bolometers (blue diamonds) showed most competitive exclusion limits. More recently, the development of liquid noble-gas detectors (red and green triangles) made it possible to significantly increase the target masses by keeping the background sufficiently low. The black line represents the level at which coherent neutrino scattering limits the WIMP sensitivity. While for 5 GeV/c^2 solar ^{8}B-neutrinos would be the first to undergo coherent neutrino scattering, for 50 GeV/c^2 atmospheric and the diffuse background of supernova neutrinos contribute.

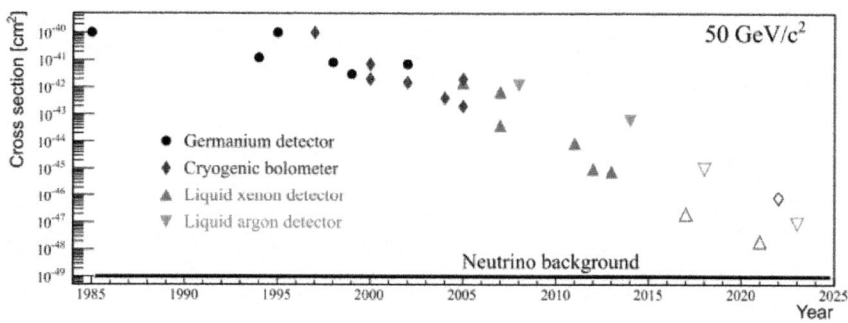

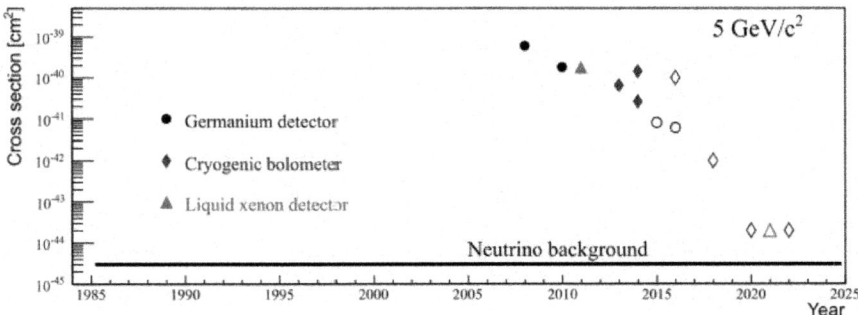

Figure 17: Evolution of sensitivity for spin-independent WIMP–nucleon cross-section for a 50 GeV/c² WIMP mass (top) and a 5 GeV/c² dark matter particle (bottom). Data from germanium and silicon detectors are shown by black circles, cryogenic bolometers by blue diamonds, liquid xenon detector by red triangles (upwards) and liquid argon by green triangles (downwards). Empty markers represent the planned sensitivity for each technology. Below the horizontal line, the sensitivity to discover dark matter is limited by coherent neutrino scattering.

The next generation of experiments (empty markers) will aim at enlarged target masses to achieve an even higher sensitivity. At the same time, it is essential to simultaneously reduce backgrounds from natural radioactivity and of cosmogenic origin (see section 4). For this purpose, experiments are placed deep underground inside efficient shields and use techniques such as careful screening of the materials, cleaning procedures, coating or etching the surfaces, etc, to keep the background at a minimum level. As, in general, an increase of target mass is easier for liquid noble-gas detectors, we expect that this technology will continue leading the sensitivity at large WIMP masses (above ~10 GeV/c²). There is, indeed, a number of proposed detectors aiming to reach cross-section sensitivities down to ~10^{-46} cm². The DEAP3600 [303] and XENON1T [324] are ton-scale experiments which are expected to start taking data in 2015. As a next step, detectors with target masses of several tons as DarkSide [314], LZ [325], XMASS2 [311] and XENONnT [160] are planned. The success of the liquid noble-gas TPC technology has motivated proposals for even larger detectors like the DARWIN (dark matter WIMP search with noble liquids) facility [363] in Europe, consisting of large LXe and LAr detectors. In the case of evidence of a dark matter signal DARWIN, with a total mass of about 20 tons, could make a high statistics measurement of the dark matter particle properties, i.e. its mass and cross-section. For an exposure of 200 ton × y, SI cross sections as low as 2.5×10^{-49} cm² can be tested for WIMP masses around 40 GeV/c² [364]. However, at this sensitivity the neutrino background becomes a significant background.

For WIMP masses below 10 GeV/c^2, cryogenic bolometers and new developments such as CCD cameras [346] feature best sensitivities. Therefore, instead of focusing on increasing the mass to the ton scale, these technologies have started to develop ideas in order to reach lower energy thresholds and improved background levels. The SuperCDMS [288] and the EURECA [297] experiments aim to operate a few hundreds of kg of target material to cover a significant part of the parameter space at low WIMP masses. As shown in figure 15, the DAMIC experiment is also sensitive to the currently lowest detectable WIMP masses. A future run using 100 g target material expects to improve the current sensitivity by more than two orders of magnitude [365].

The future projects mentioned above will be challenged by the requirement of reducing the external and internal backgrounds to lowest levels. However, the experiments are entering into a cross-section region in which the background from neutrinos can no longer be neglected. Neutrinos can produce both electronic recoils from their interactions with electrons and nuclear recoils from coherent neutrino scattering. Solar neutrinos interacting coherently with nuclei will start limiting the sensitivity of dark matter experiments for low WIMP masses (few GeV/c^2) for cross-sections around $\sim 10^{-45}$ cm^2. For experiments with larger energy thresholds, the coherent scattering of atmospheric neutrinos will limit the sensitivity for dark matter searches at cross-sections of $\sim 10^{-49}$ cm^2 [165, 167]. Although there are strategies to overcome the neutrino background at these cross-sections [366], ideally, a dark matter discovery would appear before neutrinos become a challenging background. Such a measurement would provide information on one of the most important topics of modern physics.

ACKNOWLEDGMENTS

We gratefully acknowledge the support by the Max-Planck society and the DFG research training group 'Particle physics beyond the standard model'. We thank our colleagues Jan Conrad, Franz von Feilitzsch, Steffen Hagstotz, Jacob Lamblin, Thomas Schwetz-Mangold, Hardy Simgen, Quirin Weitzel and Michael Willers for useful comments to this document. We would also like to thank Francis Froborg, Dongming Mei and D'Ann Barker for providing numerical data for figures in section 6.

REFERENCES

1.   Hinshaw G et al (WMAP Collaboration) 2013 Nine-year Wilkinson microwave anisotropy probe (WMAP) observations: cosmological parameter results Astrophys. J. Suppl. 208 19

2.   Ade P et al (Planck Collaboration) 2014 Planck 2013 results: I. Overview of products and scientific results Astron. Astrophys. 571 A1

3.   Riess A G et al (Supernova Search Collaboration) 1998 Observational evidence from supernovae for an accelerating Universe and a cosmological constant Astron. J. 116 1009

4.   Planck Collaboration 2015 Planck 2015 results: XIII. Cosmological parameters (arXiv:1502.01589)

5.   Springel V, Frenk C S and White S D 2006 The large-scale structure of the Universe Nature 440 1137

6.   Springel V et al 2005 Simulating the joint evolution of quasars, galaxies and their large-scale distribution Nature 435 629

7.   Schaye J O 2015 The EAGLE project: simulating the evolution and assembly of galaxies and their environments Mon. Not. R. Astron. Soc. 446 521

8.   Vogelsberger M et al 2014 Introducing the illustris project: simulating the coevolution of dark and visible matter in the Universe Mon. Not. R. Astron. Soc. 444 1518

9.   Colless M et al (2dFGRS Collaboration) 2003 The 2dF Galaxy redshift survey: final data release (arXiv:astro-ph/0306581)

10.   Anderson L et al (BOSS Collaboration) 2014 The clustering of galaxies in the SDSS-III baryon oscillation spectroscopic survey: baryon acoustic oscillations in the data releases 10 and 11 Galaxy samples Mon. Not. R. Astron. Soc. 441 24

11.   Sanchez A G, Crocce M, Cabré A, Baugh C M and Gaztañaga E 2009 Cosmological parameter constraints from SDSS luminous red galaxies: a new treatment of large-scale clustering Mon. Not. R. Astron. Soc. 400 16043

12.   Cen R-y, Miralda-Escude J, Ostriker J P and Rauch M 1994 Gravitational collapse of small scale structure as the origin of the Lyman alpha forest Astrophys. J. 437 L9

13.   Lee K-G et al 2014 Lyα forest tomography from background galaxies: the first megaparsecresolution large-scale structure map at $z > 2$ Astrophys. J. 795 L12

14.   Van Waerbeke L, Mellier Y and Hoekstra H 2005 Dealing with systematics in cosmic shear studies: new results from the VIRMOS-descart survey Astron. Astrophys. 429 75

15.   Bartelmann M and Schneider P 2001 Weak gravitational lensing Phys. Rep. 340 291

16. Pontzen A and Governato F 2014 Cold dark matter heats up Nature 506 171

17. Einstein A 1936 Lens-like action of a star by the deviation of light in the gravitational field Science 84 506

18. Zwicky F 1937 Nebulae as gravitational lenses Phys. Rev. 51 290

19. Clowe D et al 2006 A direct empirical proof of the existence of dark matter Astrophys. J. 648 L109

20. Bradac M et al 2008 Revealing the properties of dark matter in the merging cluster MACSJ0025.4-1222 Astrophys. J. 687 959

21. Dawson W A et al 2012 Discovery of a dissociative Galaxy cluster merger with large physical separation Astrophys. J. 747 L42

22. Harvey D, Massey R, Kitching T, Taylor A and Tittley E 2015 The non-gravitational interactions of dark matter in colliding Galaxy clusters Science 347 1462–5

23. Kahlhoefer F, Schmidt-Hoberg K, Frandsen M T and Sarkar S 2014 Colliding clusters and dark matter self-interactions Mon. Not. R. Astron. Soc. 437 2865

24. Kapteyn J 1922 First attempt at a theory of the arrangement and motion of the sidereal system Astrophys. J. 55 302

25. Zwicky F 1933 Die rotverschiebung von extragalaktischen nebeln Helv. Phys. Acta 6 110

26. Rubin V C, Thonnard N and Ford J W K 1978 Extended rotation curves of high-luminosity spiral galaxies Astrophys. J 225 L107

27. Richards E E et al 2015 Baryonic distributions in the dark matter halo of NGC 5005 Mon. Not. R. Astron. Soc. 449 3981

28. Milgrom M 1983 A Modification of the Newtonian dynamics as a possible alternative to the hidden mass hypothesis Astrophys. J. 270 365

29. Bekenstein J D 2004 Relativistic gravitation theory for the MOND paradigm Phys. Rev. D 70 083509

30. Felten J 1984 Milgrom☐s revision of Newton☐s laws—dynamical and cosmological consequences Astrophys. J. 286 3

31. Seifert M D 2007 Stability of spherically symmetric solutions in modified theories of gravity Phys. Rev. D 76 064002

32. Mavromatos N E, Sakellariadou M and Yusaf M F 2009 Can the relativistic field theory version of modified Newtonian dynamics avoid dark matter on galactic scales? Phys. Rev. D 79 081301

33. Paczynski B 1986 Gravitational microlensing by the galactic halo Astrophys. J. 304 1

34. Alcock C et al (MACHO Collaboration) 2000 The MACHO project: microlensing results from 5.7 years of LMC observations Astrophys. J. 542 281

35. Iocco F, Mangano G, Miele G, Pisanti O and Serpico P D 2009 Primordial nucleosynthesis: from precision cosmology to fundamental physics Phys. Rep. 472 1

36. White S D, Frenk C and Davis M 1983 Clustering in a neutrino dominated Universe Astrophys. J. 274 L1

37. Tremaine S and Gunn J E 1979 Dynamical role of light neutral leptons in cosmology Phys. Rev. Lett. 42 407

38. Abazajian K N et al 2012 Light sterile neutrinos: a white paper (arXiv:1204.5379)

39. Kusenko A 2009 Sterile neutrinos: the dark side of the light fermions Phys. Rep. 481 1

40. Boyarsky A, Ruchayskiy O and Shaposhnikov M 2009 The role of sterile neutrinos in cosmology and astrophysics Annu. Rev. Nucl. Part. Sci. 59 191

41. Abazajian K, Fuller G M and Tucker W H 2001 Direct detection of warm dark matter in the x-ray Astrophys. J. 562 593

42. Gelmini G and Gondolo P 2010 DM production mechanisms (arXiv:1009.3690)

43. Jungman G, Kamionkowski M and Griest K 1996 Supersymmetric dark matter Phys. Rep. 267 195

44. Kaluza T 1921 Zum unitätsproblem in der physik Sitzungsber. Preuss. Akad. Wiss. Berlin. (Math. Phys.) July-Dec 1921, p 966

45. Klein O 1926 Quantum theory and five-dimensional theory of relativity Z. Phys. 37 895

46. Kuzmin V and Tkachev I 1998 Ultrahigh-energy cosmic rays, superheavy long living particles, and matter creation after inflation JETP Lett. 68 271

47. Greisen K 1966 End to the cosmic ray spectrum? Phys. Rev. Lett. 16 748

48. Zatsepin G and Kuzmin V 1966 Upper limit of the spectrum of cosmic rays JETP Lett. 4 78

49. Chung D J, Kolb E W and Riotto A 1999 Superheavy dark matter Phys. Rev. D 59 023501

50. Baker C et al 2006 An improved experimental limit on the electric dipole moment of the neutron Phys. Rev. Lett. 97 131801

51. Raffelt G G 2008 Astrophysical axion bounds Lecture Notes Phys. 741 51

52. Peccei R and Quinn H R 1977 CP conservation in the presence of instantons Phys. Rev. Lett. 38 1440

53. Weinberg S 1978 A new light boson? Phys. Rev. Lett. 40 223

54. Kim J E 1979 Weak interaction singlet and strong CP invariance Phys. Rev. Lett. 43 103

55. Shifman M A, Vainshtein A and Zakharov V I 1980 Can confinement ensure natural CP invariance of strong interactions? Nucl. Phys. B 166 493

56. Zhitnitsky A 1980 On possible suppression of the axion hadron interactions Sov. J. Nucl. Phys. 31 260 (in Russian)

57. Dine M, Fischler W and Srednicki M 1981 A simple solution to the strong CP problem with a harmless axion Phys. Lett. B 104 199

58. Sikivie P 1983 Experimental tests of the invisible axion Phys. Rev. Lett. 51 1415

59. Abbott L and Sikivie P 1983 A cosmological bound on the invisible axion Phys. Lett. B 120 133

60. Visinelli L and Gondolo P 2009 Dark matter axions revisited Phys. Rev. D 80 035024

61. Bertone G (ed) 2010 Particle Dark Matter (Cambridge: Cambridge University Press)

62. Chatrchyan S et al (CMS Collaboration) 2008 The CMS experiment at the CERN LHC JINST 3 S08004

63. Aad G et al (ATLAS Collaboration) 2008 The ATLAS experiment at the CERN large hadron collider JINST 3 S08003

64. Chatrchyan S et al (CMS Collaboration) 2012 Observation of a new boson at a mass of 125 GeV with the CMS experiment at the LHC Phys. Lett. B 716 30

65. Aad G et al (ATLAS Collaboration) 2012 Observation of a new particle in the search for the standard model Higgs boson with the ATLAS detector at the LHC Phys. Lett. B 716 1

66. Aad G et al (ATLAS Collaboration) 2014 Search for dark matter in events with a hadronically decaying W or Z boson and missing transverse momentum in pp collisions at s = 8 TeV with the ATLAS detector Phys. Rev. Lett. 112 041802

67. Chatrchyan S et al CMS Collaboration 2012 Search for dark matter and large extra dimensions in monojet events in pp collisions at s = 7 TeV J. High Energy Phys. JHEP09(2012)094

68. Aad G et al (ATLAS Collaboration) 2013 Search for dark matter candidates and large extra dimensions in events with a photon and missing transverse momentum in pp collision data at s = 7 TeV with the ATLAS detector Phys. Rev. Lett. 110 011802

69. Strigari L E 2013 Galactic searches for dark matter Phys. Rep. 531 1

70. Ibarra A, Tran D and Weniger C 2013 Indirect searches for decaying dark matter Int. J. Mod. Phys. A 28 1330040

71. Aleksic J et al (MAGIC Collaboration) 2011 Searches for dark matter annihilation signatures in the segue 1 satellite Galaxy with the MAGIC-I telescope J. Cosmol. Astropart. Phys. JCAP06 (2011)035

72. Rico J, Wood M, Drlica-Wagner A, Aleksi J and (Fermi-LAT, MAGIC Collaboration) 2015 Limits to dark matter properties from a combined analysis of MAGIC and Fermi-LAT observations of dwarf satellite galaxies Proc. 34th Int. Cosmic Ray Conf. (ICRC 2015) (arXiv:1508.05827)

73. Abramowski A et al (H.E.S.S. Collaboration) 2013 Search for photon line-like signatures from dark matter annihilations with H.E.S.S Phys. Rev. Lett. 110 041301

74. Abramowski A et al (H.E.S.S. Collaboration) 2015 Constraints on an annihilation signal from a core of constant dark matter density around the Milky Way center with H.E.S.S Phys. Rev. Lett. 114 081301

75. Arlen T et al (VERITAS Collaboration) 2012 Constraints on cosmic rays, magnetic fields, and dark matter from gamma-ray observations of the coma cluster of galaxies with VERITAS and Fermi Astrophys. J. 757 123

76. Zitzer B and (VERITAS Collaboration) 2015 The VERITAS dark matter program 5th Int. Fermi Symp. (Nagoya, Japan, 20–24 October 2014) (arXiv:1503.00743)

77. Conrad J, Cohen-Tanugi J and Strigari L E 2015 WIMP searches with gamma rays in the Fermi era: challenges, methods and results JETP 148 12

78. Ackermann M et al (Fermi-LAT Collaboration) 2014 The spectrum and morphology of the Fermi bubbles Astrophys. J. 793 64

79. Ackermann M et al (Fermi-LAT Collaboration) 2013 Search for gamma-ray spectral lines with the Fermi large area telescope and dark matter implications Phys. Rev. D 88 082002

80. Hooper D and Goodenough L 2011 Dark matter annihilation in the Galactic center as seen by the Fermi gamma ray space telescope Phys. Lett. B 697 412

81. Ackermann M et al (Fermi-LAT Collaboration) 2015 Limits on dark matter annihilation signals from the Fermi LAT 4-year measurement of the isotropic gamma-ray background (arXiv:1501.05464)

82. Ackermann M et al (Fermi-LAT Collaboration) 2015 Searching for dark matter annihilation from Milky Way dwarf spheroidal galaxies with six years of Fermi-LAT data (arXiv:1503.02641)

83. Bulbul E et al 2014 Detection of an unidentified emission line in the stacked x-ray spectrum of Galaxy clusters Astrophys. J. 789 13

84. Boyarsky A, Ruchayskiy O, Iakubovskyi D and Franse J 2014 Unidentified line in x-ray spectra of the andromeda galaxy and perseus Galaxy cluster Phys. Rev. Lett. 113 251301

85. Abazajian K N 2014 Resonantly produced 7 keV sterile neutrino dark matter models and the properties of Milky Way satellites Phys. Rev. Lett. 112 161303

86. Ishida H, Jeong K S and Takahashi F 2014 7 keV sterile neutrino dark matter from split flavor mechanism Phys. Lett. B 732 196

87. Higaki T, Jeong K S and Takahashi F 2014 The 7 keV axion dark matter and the x-ray line signal Phys. Lett. B 733 25

88. Jaeckel J, Redondo J and Ringwald A 2014 3.55 keV hint for decaying axionlike particle dark matter Phys. Rev. D 89 103511

89. Jeltema T E and Profumo S 2015 Discovery of a 3.5 keV line in the Galactic center and a critical look at the origin of the line across astronomical targets Mon. Not. R. Astron. Soc. 450 2143

90. Carlson E, Jeltema T and Profumo S 2015 Where do the 3.5 keV photons come from? A morphological study of the Galactic center and of perseus J. Cosmol. Astropart. Phys. JCAP02 (2015)009

91. Bulbul E et al 2014 Comment on 'dark matter searches going bananas: the contribution of potassium (and chlorine) to the 3.5 keV line' (arXiv:1409.4143)

92. Aartsen M et al (IceCube Collaboration) 2013 Search for dark matter annihilations in the Sun with the 79-string IceCube detector Phys. Rev. Lett. 110 131302

93. Choi K et al (Super-Kamiokande Collaboration) 2015 Search for neutrinos from annihilation of captured low-mass dark matter particles in the Sun by Super-Kamiokande Phys. Rev. Lett. 114 141301

94. Adriani O et al (PAMELA Collaboration) 2009 An anomalous positron abundance in cosmic rays with energies 1.5–100 GeV Nature 458 607

95. Aguilar M et al (AMS Collaboration) 2013 First result from the alpha magnetic spectrometer on the international space station: precision measurement of the positron fraction in primary cosmic rays of 0.5350 GeV Phys. Rev. Lett. 110 141102

96. Blum K, Katz B and Waxman E 2013 AMS-02 results support the secondary origin of cosmic ray positrons Phys. Rev. Lett. 111 211101

97. Lewin J and Smith P 1996 Review of mathematics, numerical factors, and corrections for dark matter experiments based on elastic nuclear recoil Astropart. Phys. 6 87

98. Goodman M W and Witten E 1985 Detectability of certain dark matter candidates Phys. Rev. D 31 3059

99. Drukier A, Freese K and Spergel D 1986 Detecting cold dark matter candidates Phys. Rev. D 33 3495

100. Freese K, Lisanti M and Savage C 2013 Colloquium: annual modulation of dark matter Rev. Mod. Phys. 85 1561

101. Spergel D N 1988 The motion of the Earth and the detection of weakly interacting massive particles Phys. Rev. D 37 1353

102. Snowden-Ifft D P, Martoff C J and Burwell J M 2000 Low pressure negative ion drift chamber for dark matter search Phys. Rev. D 61 101301

103. Helm R H 1956 Inelastic and elastic scattering of 187 MeV electrons from selected even–even nuclei Phys. Rev. 104 1466

104. Vietze L, Klos P, Menndez J, Haxton W and Schwenk A 2015 Nuclear structure aspects of spinindependent WIMP scattering off xenon Phys. Rev. D 91 043520

105. Ressell M and Dean D 1997 Spin dependent neutralino—nucleus scattering for A approximately 127 nuclei Phys. Rev. C 56 535

106. Toivanen P, Kortelainen M, Suhonen J and Toivanen J 2009 Large-scale shell-model calculations of elastic and inelastic scattering rates of lightest supersymmetric particles (LSP) on I-127, Xe- 129, Xe-131, and Cs-133 nuclei Phys. Rev. C 79 044302

107. Klos P, Menéndez J, Gazit D and Schwenk A 2013 Large-scale nuclear structure calculations for spin-dependent WIMP scattering with chiral effective field theory currents Phys. Rev. D 88 083516

108. Menéndez J, Gazit D and Schwenk A 2012 Spin-dependent WIMP scattering off nuclei Phys. Rev. D 86 103511

109. Fitzpatrick A L, Haxton W, Katz E, Lubbers N and Xu Y 2012 Model independent direct detection analyses (arXiv:1211.2818)

110. Anand N, Fitzpatrick A L and Haxton W 2014 Weakly interacting massive particle-nucleus elastic scattering response Phys. Rev. C 89 065501

111. Fitzpatrick A L, Haxton W, Katz E, Lubbers N and Xu Y 2013 The effective field theory of dark matter direct detection J. Cosmol. Astropart. Phys. JCAP02(2013)004

112. Schneck K et al (SuperCDMS Collaboration) 2015 Dark matter effective field theory scattering in direct detection experiments Phys. Rev. D 91 092004

113. Catena R and Gondolo P 2015 Global limits and interference patterns in dark matter direct detection (arXiv:1504.06554)

114. Tucker-Smith D and Weiner N 2001 Inelastic dark matter Phys. Rev. D 64 043502

115. Akimov D Y et al (ZEPLIN-III Collaboration) 2010 Limits on inelastic dark matter from ZEPLIN-III Phys. Lett. B 692 180

116. Arrenberg S and (CDMS Collaboration) 2011 Search for inelastic dark matter with the CDMS experiment PoS IDM2010 021

117. Aprile E et al (XENON100 Collaboration) 2011 Implications on inelastic dark matter from 100 live days of XENON100 data Phys. Rev. D 84 061101

118. Ellis J R, Flores R and Lewin J 1988 Rates for inelastic nuclear excitation by dark matter particles Phys. Lett. B 212 375

119. Baudis L et al 2013 Signatures of dark matter scattering inelastically off nuclei Phys. Rev. D 88 115014

120. Uchida H et al (XMASS-I Collaboration) 2014 Search for inelastic WIMP nucleus scattering on 129Xe in data from the XMASS-I experiment Prog. Theor. Exp. Phys. 2014 063C01

121. Essig R, Mardon J and Volansky T 2012 Direct detection of sub-GeV dark matter Phys. Rev. D 85 076007

122. Essig R, Manalaysay A, Mardon J, Sorensen P and Volansky T 2012 First direct detection limits on sub-GeV dark matter from XENON10 Phys. Rev. Lett. 109 021301

123. Kopp J, Niro V, Schwetz T and Zupan J 2009 DAMA/LIBRA and leptonically interacting dark matter Phys. Rev. D 80 083502

124. Foot R 2014 A dark matter scaling relation from mirror dark matter Phys. Dark Univ. 5-6 236

125. Feldstein B, Graham P W and Rajendran S 2010 Luminous dark matter Phys. Rev. D 82 075019

126. Bernabei R et al 2006 Investigating pseudoscalar and scalar dark matter Int. J. Mod. Phys. A 21 1445

127. Abe K et al 2013 Search for solar axions in XMASS, a large liquid-xenon detector Phys. Lett. B 724 46

128. Ahmed Z et al (CDMS Collaboration) 2009 Search for axions with the CDMS experiment Phys. Rev. Lett. 103 141802

129. Armengaud E et al (EDELWEISS-II Collaboration) 2013 Axion searches with the EDELWEISSII experiment J. Cosmol. Astropart. Phys. JCAP11(2013)067

130. Aprile E et al (XENON100 Collaboration) 2014 First axion results from the XENON100 experiment Phys. Rev. D 90 062009

131. Pospelov M, Ritz A and Voloshin M B 2008 Bosonic super-WIMPs as keV-scale dark matter Phys. Rev. D 78 115012

132. Abe K et al (XMASS Collaboration) 2014 Search for bosonic superweakly interacting massive dark matter particles with the XMASS-I detector Phys. Rev. Lett. 113 121301

133. Green A M 2012 Astrophysical uncertainties on direct detection experiments Mod. Phys. Lett. A 27 1230004

134. Read J 2014 The local dark matter density J. Phys. G: Nucl. Part. Phys. 41 063101

135. Kerr F J and Lynden-Bell D 1986 Review of galactic constants Mon. Not. R. Astron. Soc. 221 1023

136. Smith M C et al 2007 RAVE survey: constraining the local galactic escape speed Mon. Not. R. Astron. Soc. 379 755

137. King I R 1966 The structure of star clusters:III. Some simple dynamical models Astron. J. 71 64

138. Chaudhury S, Bhattacharjee P and Cowsik R 2010 Direct detection of WIMPs : implications of a self-consistent truncated isothermal model of the Milky Way□s dark matter halo J. Cosmol. Astropart. Phys. JCAP09(2010)020

139. Evans N W, Carollo C M and de Zeeuw P T 2000 Triaxial haloes and particle dark matter detection Mon. Not. R. Astron. Soc. 318 1131–43

140. Vogelsberger M et al 2009 Phase-space structure in the local dark matter distribution and its signature in direct detection experiments Mon. Not. R. Astron. Soc. 395 797

141. Navarro J F, Frenk C S and White S D 1996 The structure of cold dark matter halos Astrophys. J. 462 563

142. Springel V et al 2008 The aquarius project: the subhalos of galactic halos Mon. Not. R. Astron. Soc. 391 16085

143. Stadel J et al 2009 Quantifying the heart of darkness with GHALO—a multi-billion particle simulation of our galactic halo Mon. Not. R. Astron. Soc. 398 L21

144. Diemand J et al 2008 Clumps and streams in the local dark matter distribution Nature 454 735

145. Weinberg D H, Bullock J S, Governato F, de Naray R K and Peter A H G 2013 Cold dark matter: controversies on small scales Sackler Coll.: Dark Matter Universe: On the Threshold of Discovery (Irvine, CA, 18–20 Octobar 2012) (arXiv:1306.0913

146. astro-ph.CO.)

147. Spergel D N and Steinhardt P J 2000 Observational evidence for selfinteracting cold dark matter Phys. Rev. Lett. 84 3760

148. Navarro J F, Eke V R and Frenk C S 1996 The cores of dwarf Galaxy halos Mon. Not. R. Astron. Soc. 283 L72

149. Leo W R 1994 Techniques for Nuclear and Particle Physics Experiments (Berlin: Springer) 2nd revised edn

150. Heusser G 2005 Low level counting from meteorites to neutrinos AIP Conf. Proc. 785 39

151. Armengaud E et al (EDELWEISS Collaboration) 2013 Background studies for the EDELWEISS dark matter experiment Astropart. Phys. 47 1

152. Mei D and Hime A 2006 Muon-induced background study for underground laboratories Phys. Rev. D 73 053004

153. Esch E-I et al 2005 The cosmic ray muon flux at WIPP Nucl. Instrum. Methods A538 516–25

154. Waysand G et al 2000 First characterization of the ultrashielded chamber in the low noise underground laboratory (LSBB) of Rustrel Pays d'Apt Nucl. Instrum. Methods A 444 336

155. Kim S-C and (KIMS Collaboration) 2012 The recent results from KIMS experiment J. Phys.: Conf. Ser. 384 012020

156. Berger C et al (FREJUS Collaboration) 1989 Experimental study of muon bundles observed in the Frejus detector Phys. Rev. D 40 2163

157. Yu-Cheng W et al 2013 Measurement of cosmic ray flux in the China jinping underground laboratory Chin. Phys. C 37 086001

158. Aprile E et al (XENON100 Collaboration) 2012 The XENON100 dark matter experiment Astropart. Phys. 35 573

159. Akerib D et al 2004 Installation and commissioning of the CDMSII experiment at Soudan Nucl. Instrum. Methods A 520 116

160. Akerib D et al (LUX Collaboration) 2013 The large underground xenon (LUX) experiment Nucl. Instrum. Methods A 704 111–26

161. Aprile E et al (XENON1T Collaboration) 2014 Conceptual design and simulation of a water Cherenkov muon veto for the XENON1T experiment J. Instrum. 9 11006

162. Bossa M et al (DarkSide Collaboration) 2014 DarkSide-50, a background free experiment for dark matter searches J. Instrum. 9 C01034

163. Cabrera B, Krauss L M and Wilczek F 1985 Bolometric detection of neutrinos Phys. Rev. Lett. 55 25

164. Bellini G et al (BOREXINO Collaboration) 2014 Neutrinos from the primary protonproton fusion process in the Sun Nature 512 383

165. Freedman D Z 1974 Coherent neutrino nucleus scattering as a probe of the weak neutral current Phys. Rev. D 9 1389

166. Strigari L E 2009 Neutrino coherent scattering rates at direct dark matter detectors New J. Phys. 11 105011

167. Gutlein A et al 2010 Solar and atmospheric neutrinos: background sources for the direct dark matter search Astropart. Phys. 34 90

168. Ruppin F, Billard J, Figueroa-Feliciano E and Strigari L 2014 Complementarity of dark matter detectors in light of the neutrino background Phys. Rev. D 90 083510

169. Aartsen M et al (IceCube Collaboration) 2013 Seasonal variation of atmospheric neutrinos in IceCube Proc. ICRC

170. Muenster A et al 2014 Radiopurity of CaWO4 crystals for direct dark matter search with CRESST and EURECA J. Cosmol. Astropart. Phys. JCAP05(2014)018

171. Danevich F et al 2011 Effect of recrystallisation on the radioactive contamination of cawo4 crystal scintillators Nucl. Instrum. Methods A 631 44

172. Shields E, Xu J and Calaprice F 2015 SABRE: a new NaI(Tl) dark matter direct detection experiment Phys. Procedia 61 169

173. Bernabei R 2014 Crystal scintillators for low background measurements Eur. Phys. J. Web Conf. vol 65 p 01001

174. Agnese R et al (SuperCDMS Collaboration) 2013 Demonstration of surface electron rejection with interleaved germanium detectors for dark matter searches Appl. Phys. Lett. 103 164105

175. Broniatowski A et al 2009 A new high-background-rejection dark matter Ge cryogenic detector Phys. Lett. B 681 305

176. Strauss R et al (CRESST Collaboration) 2014 A detector module with highly efficient surfacealpha event rejection operated in CRESST-II phase 2 (arXiv:1410.1753)

177. Cebrian S et al 2010 Cosmogenic activation in germanium and copper for rare event searches Astropart. Phys. 33 316

178. Martoff C and Lewin P 1992 COSMO—a program to estimate spallation radioactivity produced in a pure substance by exposure to cosmic radiation on the Earth Comput. Phys. Commun. 72 96

179. Back J and Ramachers Y A 2008 ACTIVIA: calculation of isotope production cross-sections and yields Nucl. Instrum. Methods A 586 2864

180. Amar J et al 2015 Cosmogenic radionuclide production in NaI(Tl) crystals J. Cosmol. Astropart. Phys. JCAP02(2015)046

181. Back H O et al 2012 First large scale production of low radioactivity argon from underground sources (arXiv:1204.6024)

182. Akerib D et al 2015 Radiogenic and muon-induced backgrounds in the LUX dark matter detector Astropart. Phys. 62 33

183. Albert J et al (EXO-200 Collaboration) 2014 Improved measurement of the 2nbb half-life of 136Xe with the EXO-200 detector Phys. Rev. C C89 015502

184. Abe K et al (XMASS Collaboration) 2009 Distillation of liquid xenon to remove krypton Astropart. Phys. 31 290

185. Lindemann S and Simgen H 2014 Krypton assay in xenon at the ppq level using a gas chromatographic system combined with a mass spectrometer Eur. Phys. J. C 74 1

186. Aprile E, Yoon T, Loose A, Goetzke L W and Zelevinsky T 2013 An atom trap trace analysis system for measuring krypton contamination in xenon dark matter detectors Rev. Sci. Instrum. 84 093105

187. Zuzel G and Simgen H 2009 High sensitivity radon emanation measurements Appl. Radiat. Isot. 67 889

188. Battat J et al 2014 Radon in the DRIFT-II directional dark matter TPC: emanation, detection and mitigation J. Instrum. 9 P11004

189. Martens K and (XMASS Collaboration) 2012 Radon removal from liquid xenon Nucl. Phys. Proc. Suppl. 229 562

190. Lindemann S 2013 Intrinsic 85Kr and 222Rn backgrounds in the XENON dark matter search PhD Thesis University of Heidelberg

191. Heusser G 1995 Low-radioactivity background techniques Ann. Rev. Nucl. Part. Sci. 45 543

192. Bernabei R et al (DAMA/LIBRA Collaboration) 2010 New results from DAMA/LIBRA Eur. Phys. J. C 67 39

193. Angloher G et al (CRESST Collaboration) 2012 Results from 730 kg days of the CRESST-II dark matter search Eur. Phys. J. C 72 19071

194. Agnese R et al (CDMS Collaboration) 2013 Silicon detector dark matter results from the final exposure of CDMS II Phys. Rev. Lett. 111 251301

195. Aalseth C et al (CoGeNT Collaboration) 2014 Search for an annual modulation in three years of CoGeNT dark matter detector data (arXiv:1401.3295)

196. Conrad J 2014 Statistical issues in astrophysical searches for particle dark matter Astropart. Phys. 62 165

197. Feldman G J and Cousins R D 1998 A unified approach to the classical statistical analysis of small signals Phys. Rev. D 57 3873

198. Conrad J, Botner O, Hallgren A and de los Heros C P 2003 Including systematic uncertainties in confidence interval construction for Poisson statistics Phys. Rev. D 67 012002

199. Rolke W A, Lopez A M and Conrad J 2005 Limits and confidence intervals in the presence of nuisance parameters Nucl. Instrum. Methods A 551 493

200. Xiao M et al (PandaX Collaboration) 2014 First dark matter search results from the PandaX-I experiment Sci. China Phys. Mech. Astron. 57 2024

201. Akimov D Y et al 2012 WIMP-nucleon cross-section results from the second science run of ZEPLIN-III Phys. Lett. B 709 14

202. Felizardo M et al (SIMPLE Collaboration) 2012 Final analysis and results of the phase II SIMPLE dark matter search Phys. Rev. Lett. 108 201302

203. Yellin S 2002 Finding an upper limit in the presence of unknown background Phys. Rev. D 66 032005

204. Agnese R et al (SuperCDMS Collaboration) 2014 Search for low-mass weakly interacting massive particles with SuperCDMS Phys. Rev. Lett. 112 241302

205. Angloher G et al (CRESST-II Collaboration) 2014 Results on low mass WIMPs using an upgraded CRESST-II detector Eur. Phys. J. C 74 3184

206. Amole C et al (PICO Collaboration) 2015 Dark matter search results from the PICO-2L C3F8 Bubble chamber Phys. Rev. Lett. 114 231302

207. Cowan G, Cranmer K, Gross E and Vitells O 2011 Asymptotic formulae for likelihood-based tests of new physics Eur. Phys. J. C 71 1554

208. Aprile E et al (XENON100 Collaboration) 2011 Likelihood approach to the first dark matter results from XENON100 Phys. Rev. D 84 052003

209. Junk T 1999 Confidence level computation for combining searches with small statistics Nucl. Instrum. Methods A 434 435

210. Read A L 2002 Presentation of search results: The CL(s) technique J. Phys. G: Nucl. Part. Phys. 28 2693

211. Aprile E et al (XENON100 Collaboration) 2013 Limits on spin-dependent WIMP-nucleon cross sections from 225 live days of XENON100 data Phys. Rev. Lett. 111 021301

212. Akerib D et al (LUX Collaboration) 2014 First results from the LUX dark matter experiment at the Sanford Underground Research Facility Phys. Rev. Lett. 112 091303

213. Agnese R et al (SuperCDMS Collaboration) 2015 Maximum likelihood analysis of low energy CDMS II germanium data Phys. Rev. D 91 052021

214. Metropolis N, Rosenbluth A, Rosenbluth M, Teller A and Teller E 1953 Equation of state calculations by fast computing machines J. Chem. Phys. 21 1087

215. Hastings W 1970 Monte Carlo sampling methods using Markov chains and their applications Biometrika 57 97

216. Arina C 2014 Bayesian analysis of multiple direct detection experiments Phys. Dark Univ. 5 1

217. Belli P et al 2000 Extending the DAMA annual modulation region by inclusion of the uncertainties in the astrophysical velocities Phys. Rev. D 61 023512

218. Fox P J, Liu J and Weiner N 2011 Integrating out astrophysical uncertainties Phys. Rev. D 83 103514

219. Fox P J, Kribs G D and Tait T M P 2011 Interpreting dark matter direct detection independently of the local velocity and density distribution Phys. Rev. D 83 034007

220. McCabe C 2010 The astrophysical uncertainties Of dark matter direct detection experiments Phys. Rev. D 82 023530

221. Frandsen M T, Kahlhoefer F, McCabe C, Sarkar S and Schmidt-Hoberg K 2012 Resolving astrophysical uncertainties in dark matter direct detection J. Cosmol. Astropart. Phys. JCAP01 (2012)024

222. Del Nobile E, Gelmini G B, Gondolo P and Huh J-H 2015 Update on the Halo-independent comparison of direct dark matter detection data Phys. Proc. 61 45

223. Blennow M, Herrero-Garcia J, Schwetz T and Vogl S 2015 Halo-independent tests of dark matter direct detection signals: local DM density LHC, and thermal freeze-out (arXiv:1505.05710)

224. Shutt T et al 1992 Simultaneous high resolution measurement of phonons and ionization created by particle interactions in a 60 g germanium crystal at 25 mK Phys. Rev. Lett. 69 3531

225. Alessandrello A et al 1997 The thermal detection efficiency for recoils induced by low-energy nuclear reactions, neutrinos or weakly interacting massive particles Phys. Lett. B 408 465

226. Simon E et al 2003 SICANE: a detector array for the measurement of nuclear recoil quenching factors using a monoenergetic neutron beam Nucl. Instrum. Methods A 507 643

227. Barker D and Mei D 2012 Germanium detector response to nuclear recoils in searching for dark matter Astropart. Phys. 38 1

228. Xu J et al 2015 Scintillation efficiency measurement of Na recoils in NaI(Tl) below the DAMA/ LIBRA energy threshold Phys. Rev. C 92 015807

229. Plante G et al 2011 New measurement of the scintillation efficiency of low-energy nuclear recoils in liquid xenon Phys. Rev. C 84 045805

230. Regenfus C et al 2012 Study of nuclear recoils in liquid argon with monoenergetic neutrons J. Phys. Conf. Ser. 375 012019

231. Creus W et al 2015 Scintillation efficiency of liquid argon in low energy neutron–argon scattering (arXiv:1504.07878)

232. Manalaysay A 2010 Towards an improved understanding of the relative scintillation efficiency of nuclear recoils in liquid xenon (arXiv:1007.3746)

233. Cao H et al (SCENE Collaboration) 2015 Measurement of scintillation and ionization yield and scintillation pulse shape from nuclear recoils in liquid argon Phys. Rev. D 91 092007

234. Alexander T et al (SCENE Collaboration) 2013 Observation of the dependence on drift field of scintillation from nuclear recoils in liquid argon Phys. Rev. D 88 092006

235. Horn M et al 2011 Nuclear recoil scintillation and ionisation yields in liquid xenon from ZEPLIN-III data Phys. Lett. B 705 471

236. Aprile E et al (XENON100 Collaboration) 2013 Response of the XENON100 dark matter detector to nuclear recoils Phys. Rev. D **88** 012006

237. Lidnhard J, Scharff M and Schiott H 1963 Range concepts and heavy ion ranges Mat. Fys. Medd. Dan. Vid. Selsk **33** 14

238. Barker D, Wei W Z, Mei D M and Zhang C 2013 Ionization efficiency study for low energy nuclear recoils in germanium (arXiv:1304.6773)

239. Mei D-M, Yin Z-B, Stonehill L and Hime A 2008 A model of nuclear recoil scintillation efficiency in noble liquids Astropart. Phys. **30** 12

240. Bezrukov F, Kahlhoefer F and Lindner M 2011 Interplay between scintillation and ionization in liquid xenon dark matter searches Astropart. Phys. **35** 119

241. Szydagis M, Fyhrie A, Thorngren D and Tripathi M 2013 Enhancement of nest capabilities for simulating low-energy recoils in liquid xenon J. Instrum. **8** C10003

242. Birks J B 1964 The Theory and Practice of Scintillation Counting (London: Pergamon)

243. Seitz F 1958 On the theory of the bubble chamber Phys. Fluids (1958–1988) **1** 2

244. Behnke E et al (COUPP Collaboration) 2012 First dark matter search results from a 4 kg CF3I Bubble chamber operated in a deep underground site Phys. Rev. D **86** 052001

245. Baudis L et al 2013 Response of liquid xenon to Compton electrons down to 1.5 keV Phys. Rev. D **87** 115015

246. Ahmed Z et al (CDMS-II Collaboration) 2010 Dark matter search results from the CDMS II experiment Science **327** 1619

247. Lebedenko V et al 2009 Result from the first science run of the ZEPLIN-III dark matter search experiment Phys. Rev. D **80** 052010

248. Benetti P et al 2008 First results from a dark matter search with liquid argon at 87 K in the Gran Sasso Underground Laboratory Astropart. Phys. **28** 495

249. Miyajima M et al 1974 Average energy expended per ion pair in liquid argon Phys. Rev. A **9** 1438

250. Takahashi T et al 1975 Average energy expended per ion pair in liquid xenon Phys. Rev. A **12** 1771

251. Bernabei R et al (DAMA Collaboration) 2008 The DAMA/LIBRA apparatus Nucl. Instrum. Methods A **592** 297

252. Bernabei R et al 2013 Final model independent result of DAMA/LIBRA-phase 1 Eur. Phys. J. C 73 2648

253. Bernabei R et al (DAMA Collaboration) 2008 First results from DAMA/LIBRA and the combined results with DAMA/NaI Eur. Phys. J. C 56 333

254. Bernabei R et al 2014 New results from DAMA/LIBRA: final model-independent results of DAMA/LIBRA-phase1 and perspectives of phase2 Frascati Phys. Ser. 58 41

255. Savage C, Gelmini G, Gondolo P and Freese K 2009 Compatibility of DAMA/LIBRA dark matter detection with other searches J. Cosmol. Astropart. Phys. JCAP04(2009)010

256. Bernabei R et al 2001 Investigating the DAMA annual modulation data in a mixed coupling framework Phys. Lett. B 509 197

257. Bernabei R et al 2002 Investigating the DAMA annual modulation data in the framework of inelastic dark matter Eur. Phys. J. C 23 61

258. Schnee R 2011 Introduction to dark matter experiments (arXiv:1101.5205)

259. Blum K 2011 DAMA versus the annually modulated muon background (arXiv:1110.0857)

260. Davis J H 2014 Fitting the annual modulation in DAMA with neutrons from muons and neutrinos Phys. Rev. Lett. 113 081302

261. Ralston J P 2010 One model explains DAMA/LIBRA, CoGENT, CDMS, and XENON (arXiv:1006.5255)

262. Barbeau P, Collar J, Efremenko Y and Scholberg K 2014 Comment on fitting the annual modulation in DAMA with neutrons from muons and neutrinos Phys. Rev. Lett. 113 229001

263. Klinger J and Kudryavtsev V 2015 Muon-induced neutrons do not explain the DAMA data Phys. Rev. Lett. 114 151301

264. Davis J H 2015 The past and future of light dark matter direct detection Int. J. Mod. Phys. A 30 1530038

265. Amaré J et al 2015 From ANAIS-25 towards ANAIS-250 Phys. Proc. 61 157

266. Cherwinka J et al (DM-Ice17 Collaboration) 2014 First data from DM-Ice17 Phys. Rev. D 90 092005

267. Kim S et al (KIMS Collaboration) 2012 New limits on interactions between weakly interacting massive particles and nucleons obtained with CsI(Tl) crystal detectors Phys. Rev. Lett. 108 181301

268. Kim K et al 2014 Tests on NaI(Tl) crystals for WIMP search at the Yangyang Underground Laboratory Astropart. Phys. 62 249

269. Radeka V 1988 Low-noise techniques in detectors Ann. Rev. Nuc. Part. Sci. 38 217

270. Ahlen S et al 1987 Limits on cold dark matter candidates from an ultralow background germanium spectrometer Phys. Lett. B 195 603

271. Aalseth C et al CoGeNT Collaboration 2013 CoGeNT: a search for low-mass dark matter using p-type point contact germanium detectors Phys. Rev. D 88 012002

272. Davis J H, McCabe C and Boehm C 2014 Quantifying the evidence for dark matter in CoGeNT data J. Cosmol. Astropart. Phys. JCAP08(2014)014

273. Aalseth C et al 2014 Maximum likelihood signal extraction method applied to 3.4 years of cogent data (arXiv:1401.6234)

274. Derevianko A, Dzuba V, Flambaum V and Pospelov M 2010 Axio-electric effect Phys. Rev. D 82 065006

275. Aalseth C et al (CoGeNT Collaboration) 2008 Experimental constraints on a dark matter origin for the DAMA annual modulation effect Phys. Rev. Lett. 101 251301

276. Bonicalzi R et al (CC-4 Collaboration) 2013 The C-4 dark matter experiment Nucl. Instrum. Methods A 712 27

277. Giovanetti G et al (Majorana Collaboration) 2014 A dark matter search with MALBEK Phys. Proc. 00 1

278. Liu S et al (CDEX and TEXONO Collaboration) 2014 Limits on light WIMPs with a germanium detector at 177 eVee threshold at the China Jinping Underground Laboratory Phys. Rev. D 90 032003

279. Lin S et al (TEXONO Collaboration) 2009 New limits on spin-independent and spin-dependent couplings of low-mass WIMP dark matter with a germanium detector at a threshold of 220 eV Phys. Rev. D 79 061101

280. Li H et al (TEXONO Collaboration) 2013 Limits on spin-independent couplings of WIMP dark matter with a p-type point-contact germanium detector Phys. Rev. Lett. 110 261301

281. Yue Q et al (CDEX Collaboration) 2014 Limits on light WIMPs from the CDEX-1 experiment with a p-type point-contact germanium detector at the China Jingping Underground Laboratory Phys. Rev. D 90 091701

282. Luke P, Beeman J, Goulding F, Labov S and Silver E 1990 Calorimetric ionization detector Nucl. Instrum. Methods A 289 406

283. Neganov B and Trofimov V 1985 Otkrytiya Izobret 146 215

284. Booth N, Cabrera B and Fiorini E 1996 Low-temperature particle detectors Ann. Rev. Nucl. Part. Sci. 46 471

285. Ahmed Z et al (CDMS Collaboration) 2009 Search for weakly interacting massive particles with the first five-tower data from the cryogenic dark matter search at the Soudan Underground Laboratory Phys. Rev. Lett. 102 011301

286. Ahmed Z et al (CDMS Collaboration) 2012 Search for annual modulation in low-energy CDMSII data (arXiv:1203.1309)

287. Agnese R et al (SuperCDMS Collaboration) 2014 search for low-mass weakly interacting massive particles using voltage-assisted calorimetric ionization detection in the SuperCDMS experiment Phys. Rev. Lett. 112 041302

288. Agnese R et al (SuperCDMS Collaboration) 2015 WIMP-search results from the second CDMSlite run (arXiv:1509.02448)

289. Cushman P et al 2013 Working Group Report: WIMP Dark Matter Direct Detection (arXiv:1310.8327)

290. Armengaud E et al (EDELWEISS Collaboration) 2011 Final results of the EDELWEISS-II WIMP search using a 4 kg array of cryogenic germanium detectors with interleaved electrodes Phys. Lett. B 702 329

291. Armengaud E et al (EDELWEISS Collaboration) 2012 A search for low-mass WIMPs with EDELWEISS-II heat-and-ionization detectors Phys. Rev. D 86 051701

292. de Boissire T and (EDELWEISS Experiment Collaboration) 2015 Low mass WIMP search with EDELWEISS-III: first results (arXiv:1504.00820)

293. Angloher G et al 2005 Limits on WIMP dark matter using scintillating CaWO4 cryogenic detectors with active background suppression Astropart. Phys. 23 325

294. Angloher G et al (CRESST Collaboration) 2015 Results on light dark matter particles with a lowthreshold CRESST-II detector (arXiv:1509.01515)

295. Angloher G et al (CRESST Collaboration) 2015 Probing low WIMP masses with the next generation of CRESST detector (arXiv:1503.08065)

296. Isaila C et al 2012 Low-temperature light detectors: Neganov–Luke amplification and calibration Phys. Lett. B 716 160

297. Cebrian S et al 2001 First results of the ROSEBUD dark matter experiment Astropart. Phys. 15 79

298. Angloher G et al (EURECA Collaboration) 2014 EURECA conceptual design report Phys. Dark Univ. 3 41

299. Lippincott W et al (MiniCLEAN Collaboration) 2012 Scintillation yield and time dependence from electronic and nuclear recoils in liquid neon Phys. Rev. C 86 015807

300. Cheshnovsky O, Raz B and Jortner J 1972 Temperature dependence of rare gas molecular emission in the vacuum ultraviolet Chem. Phys. Lett. 15 475

301. Bolozdynya A 1999 Two-phase emission detectors and their applications Nucl. Instrum. Methods A 422 314

302. Boulay M and Hime A 2006 Technique for direct detection of weakly interacting massive particles using scintillation time discrimination in liquid argon Astropart. Phys. 25 179

303. Hitachi A et al 1983 Effect of ionization density on the time dependence of luminescence from liquid argon and xenon Phys. Rev. B 27 5279

304. Boulay M and (DEAP Collaboration) 2012 DEAP-3600 dark matter search at SNOLAB J. Phys. Conf. Ser. 375 012027

305. Rielage K et al (MINICLEAN Collaboration) 2015 Update on the MiniCLEAN dark matter experiment Phys. Proc. 61 144

306. Boulay M et al 2009 Measurement of the scintillation time spectra and pulse-shape discrimination of low-energy beta and nuclear recoils in liquid argon with DEAP-1 (arXiv:0904.2930)

307. Amaudruz P-A et al 2014 Radon backgrounds in the DEAP-1 liquid argon based dark matter detector Astropart. Phys. 62 178

308. Ueshima K et al (XMASS Collaboration) 2011 Scintillation-only based pulse shape discrimination for nuclear and electron recoils in liquid xenon Nucl. Instrum. Meth. A 659 161

309. Abe K et al 2013 XMASS detector Nucl. Instrum. Methods A 716 78

310. Abe K et al (XMASS Collaboration) 2013 Light WIMP search in XMASS Phys. Lett. B 719 78

311. Liu J and (XMASS Collaboration) 2014 The XMASS experiment AIP Conf. Proc. 1604 397

312. Hiraide K et al (XMASS Collaboration) 2015 XMASS: recent results and status (arXiv:1506.08939)

313. Lansiart A, Seigneur A, Moretti J-L and Morucci J-P 1976 Development research on a highly luminous condensed xenon scintillator Nucl. Instrum. Methods 135 47

314. Agnes P et al (DarkSide Collaboration) 2015 First results from the darkside-50 dark matter experiment at Laboratori Nazionali del Gran Sasso Phys. Lett. B 743 456

315. Aalseth C E et al 2015 The darkside multiton detector for the direct dark matter search Adv. High Energy Phys. 2015 541362

316. Badertscher A et al 2013 ArDM: first results from underground commissioning JINST 8 C09005

317. Calvo J et al (ArDM Collaboration) 2015 Status of ArDM-1t: first observations from operation with a full ton-scale liquid argon target (arXiv:1505.02443)

318. Akimov D Y et al 2007 The ZEPLIN-III dark matter detector: instrument design, manufacture and commissioning Astropart. Phys. 27 46

319. Aprile E et al (XENON Collaboration) 2011 Design and performance of the XENON10 dark matter experiment Astropart. Phys. 34 679

320. Angle J et al (XENON Collaboration) 2008 First results from the XENON10 dark matter experiment at the Gran Sasso National Laboratory Phys. Rev. Lett. 100 021303

321. Angle J et al (XENON10 Collaboration) 2011 A search for light dark matter in XENON10 data Phys. Rev. Lett. 107 051301

322. Aprile E et al (XENON100 Collaboration) 2012 Dark matter results from 225 live days of XENON100 data Phys. Rev. Lett. 109 181301

323. Aprile E et al (XENON100 Collaboration) 2015 Search for event rate modulation in XENON100 electronic recoil data Phys. Rev. Lett. 115 091302

324. Aprile E et al (XENON100 Collaboration) 2015 Exclusion of leptophilic dark matter models using XENON100 electronic recoil data Science 349 851

325. Aprile E et al (XENON1T Collaboration) 2013 The XENON1T dark matter search experiment Proc. 10th UCLA Symp. on Sources and Detection of Dark Matter and Dark Energy in the Universe (Springer Proceedings in Physics 148) (New York: Springer) p 93

326. Malling D et al 2011 After LUX: the LZ program (arXiv:1110.0103)

327. LZ Collaboration Akerib D S et al 2015 LUX-ZEPLIN (LZ) Conceptual Design Report (arXiv:1509.02910)

328. Cao X et al (PandaX Collaboration) 2014 PandaX: a liquid xenon dark matter experiment at CJPL Sci. China Phys. Mech. Astron. 57 1476

329. Xiao X et al (PANDA-X Collaboration) 2015 Low-mass dark matter search results from full exposure of PandaX-I experiment Phys. Rev. D 92 052001

330. Behnke E et al (COUPP Collaboration) 2008 Spin-dependent WIMP limits from a bubble chamber Science 319 933

331. Barnabe-Heider M et al (PICASSO Collaboration) 2005 Response of superheated droplet detectors of the picasso dark matter search experiment Nucl. Instrum. Methods A 555 184

332. Archambault S et al (PICASSO Collaboration) 2012 Constraints on low-mass WIMP interactions on F19 from PICASSO Phys. Lett. B 711 153

333. Ahlen S et al 2010 The case for a directional dark matter detector and the status of current experimental efforts Int. J. Mod. Phys. A 25 1

334. Sauli F and Sharma A 1999 Micropattern gaseous detectors Ann. Rev. Nucl. Part. Sci. 49 341

335. Battat J et al (DRIFT Collaboration) 2014 First background-free limit from a directional dark matter experiment: results from a fully fiducialised DRIFT detector Phys. Dark. Univ. 9-10 1

336. Billard J et al 2014 In situ measurement of the electron drift velocity for upcoming directional dark matter detectors JINST 9 P01013

337. Giomataris Y, Rebourgeard P, Robert J and Charpak G 1996 MICROMEGAS: a high granularity position sensitive gaseous detector for high particle flux environments Nucl. Instrum. Methods A 376 29

338. Riffard Q et al 2015 First detection of tracks of radon progeny recoils by MIMAC (arXiv:1504.05865)

339. Battat J B et al 2014 The dark matter time projection chamber 4Shooter directional dark matter detector: calibration in a surface laboratory Nucl. Instrum. Meth. A 755 6

340. Deaconu C et al 2015 Track reconstruction progress from the DMTPC directional dark matter experiment Phys. Proc. 61 39

341. Miuchi K et al 2010 First underground results with NEWAGE-0.3a direction-sensitive dark matter detector Phys. Lett. B 686 11

342. Vahsen S et al 2012 The directional dark matter detector (D3) EAS Publ. Ser. 53 43

343. Nakamura K et al 2015 Direction-sensitive dark matter search with gaseous tracking detector NEWAGE-0.3b PTEP 2015 043F01

344. Ross S 2014 Recent progress on D3 —the directional dark matter detector (arXiv:1402.0043)

345. Naka T et al 2013 Fine grained nuclear emulsion for higher resolution tracking detector Nucl. Instrum. Methods A 718 519

346. Yoo J et al 2015 Scalability study of solid xenon JINST 10 P04009

347. Barreto J et al (DAMIC Collaboration) 2012 Direct search for low mass dark matter particles with CCDs Phys. Lett. B 711 264

348. Aguilar-Arevalo A et al (DAMIC Collaboration) 2015 Measurement of radioactive contamination in the high-resistivity silicon CCDs of the DAMIC experiment JINST 10 P08010

349. Gerbier G et al 2014 NEWS : a new spherical gas detector for very low mass WIMP detection (arXiv:1401.7902)

350. Iguaz F et al 2015 TREX-DM: a low background micromegas-based TPC for low mass WIMP detection (arXiv:1503.07085)

351. Bertoni R et al 2014 A new technique for direct investigation of dark matter Nucl. Instrum. Methods A 744 61

352. Drukier A et al 2012 New dark matter detectors using DNA for nanometer tracking (arXiv:1206.6809)

353. http://dmtools.brown.edu

354. Goodman J, Ibe M, Rajaraman A, Shepherd W, Tait T M P and Yu H-B 2010 Constraints on dark matter from colliders Phys. Rev. D 82 116010

355. Beltran M, Hooper D, Kolb E W, Krusberg Z A C and Tait T M P 2010 Maverick dark matter at colliders J. High Energy Phys. JHEP09(2010)037

356. Fox P J, Harnik R, Kopp J and Tsai Y 2012 Missing energy signatures of dark matter at the LHC Phys. Rev. D 85 056011

357. DiFranzo A, Nagao K I, Rajaraman A and Tait T M 2013 Simplified models for dark matter interacting with quarks J. High Energy Phys. JHEP11(2013)014

358. Buchmueller O, Dolan M J, Malik S A and McCabe C 2015 Characterising dark matter searches at colliders and direct detection experiments: vector mediators J. High Energy Phys. JHEP01 (2015)037

359. Buchmueller O, Dolan M J and McCabe C 2014 Beyond effective field theory for dark matter searches at the LHC J. High Energy Phys. JHEP01(2014)025

360. Busoni G, De Simone A, Morgante E and Riotto A 2014 On the validity of the effective field theory for dark matter searches at the LHC Phys. Lett. B 728 412

361. Press W H and Spergel D N 1985 Capture by the Sun of a galactic population of weakly interacting massive particles Astrophys. J. 296 679

362. Silk J, Olive K A and Srednicki M 1985 The photino, the Sun and high-energy neutrinos Phys. Rev. Lett. 55 257

363. Boliev M, Demidov S, Mikheyev S and Suvorova O 2013 Search for muon signal from dark matter annihilations inthe Sun with the Baksan underground scintillator telescope for 24.12 years J. Cosmol. Astropart. Phys. JCAP09(2013)019

364. Baudis L and (Darwin Collaboration) 2011 DARWIN: dark matter WIMP search with noble liquids Proc. 8th Int. Workshop on the Identification of Dark Matter (IDM 2010) (Proc. Sci.) p 122

365. Schumann M, Baudis L, Butikofer L, Kish A and Selvi M 2015 Dark matter sensitivity of multiton liquid xenon detectors J. Cosmol. Astropart. Phys. JCAP10(2015)016

366. Chavarria A E et al 2015 DAMIC at SNOLAB Phys. Proc. 61 21

367. O'Hare C A J, Green A M, Billard J, Figueroa-Feliciano E and Strigari L E 2015 Readout strategies for directional dark matter detection beyond the neutrino background Phys. Rev. D 92 063518

Chapter 6

THE USE OF AVALANCHE PHOTODIODES IN HIGH ENERGY ELECTROMAGNETIC CALORIMETRY

Paola La Rocca[1,2] and Francesco Riggi[2]

[1]Museo Storico della Fisica e Centro Studi e Ricerche "E.Fermi"

[2]Department of Physics and Astronomy, University of Catania Italy

INTRODUCTION

Avalanche Photodiodes (APD) are now widely used for the detection of weak optical signals. They find applications in a large number of fields of science and technology, from physics to medicine and environmental sciences. The request for sensitive detectors, capable to respond to weak radiations emitted from scintillation materials, has produced over the last decades an increasing number of studies on avalanche photodiodes and their applications as photo-sensors in particle detection. The first APD prototypes were developed more than 40 years ago. The initial size of such devices was however very small (below 1 mm^2) and their spectral response confined to the near-infrared region. As a result, although available since several years, they did not receive much attention, also because of their initial high cost and low gain. However, large progresses have been made since then, and it has been possible to design and produce, at a reasonable cost, devices which have now a much larger area (tens of squared millimetres), with a high spectral sensitivity in the blue and near ultra-violet wavelength region. For such reasons, avalanche photodiodes are now widely used as sensitive light detectors in the construction of particle detectors in high energy physics. One of such examples is the impulse received by the design and construction of large scale electromagnetic calorimeters for the high energy experiments currently running in the world largest Laboratories. At present, APDs exhibit excellent quantum efficiency, with values around 80% in the near ultra-violet range, dropping to about 40% in the blue region, which is to be compared to typical values of 5-8 % in the blue for standard photomultipliers. Additional advantages which make them preferable over photomultipliers are discussed more specifically in the Chapter.

The overall set of problems and solutions related to the use of Avalanche Photodiodes in the design, construction, test and operation of large electromagnetic calorimeters in nuclear and particle physics experiments, is described in this Chapter, as observed within a Collaboration at the CERN Large Hadron Collider. Section 2 briefly recalls the principles on which electromagnetic calorimetry for particle physics experiments is based. Relative merits of Avalanche Photodiodes in comparison to traditional devices, mostly photomultipliers, are discussed in Sect.3, in connection with the light collection from scintillation detectors and the readout and front-end electronics. A review of the large detectors which have employed in the recent past or are currently employing such devices as photo-sensors is given in Section 4. Sect.5 describes the overall set of procedures carried out to characterize a large number of such devices when installing a complex detector. Section 6 discusses also the problems which may be encountered in the digital treatment of the signal and presents a comparison between traditional and alternative approaches in the analysis procedures.

HIGH ENERGY ELECTROMAGNETIC CALORIMETRY IN NUCLEAR AND PARTICLE PHYSICS

The use of avalanche photodiodes in nuclear and particle physics has largely increased in the last decades especially in connection with the growing impact of calorimetry techniques on accelerator-based physics experimentation being taken in the world largest Laboratories. The term "calorimetry" comes from the Latin word calor (= heat) and indicates the basic detection principle which calorimeters are based on: the incident particles to be measured are fully absorbed in a block of instrumented material and their energy is converted into a measurable quantity (usually charge or light). In the process of absorption showers of secondary particles are generated, causing a progressive degradation in energy and producing some signal which can be detected to gain information on the original energy of the particle.

In order to match the physics potential at the major particle accelerator facilities, a wide variety of possible solutions for calorimeters is today available. Apart from the broad distinction between electromagnetic and hadronic calorimeters, they can be further classified according to the various types of technology employed, sampling calorimeters and homogeneous calorimeters being the most commonly used. This Chapter will focus on the electromagnetic calorimetry, describing the working principle and the practical realizations of electromagnetic calorimeters, as well as the reasons that make such detectors so attractive in the field of nuclear and particle physics. The interested reader is referred to textbooks (Wigmans, 2000) or review papers (Fabjan, 2003) for a more detailed discussion on calorimeters.

Working principle of an Electromagnetic Calorimeter

The various interaction mechanisms by which particles of different nature lose their energy in the medium underlie the broad distinction between hadronic and electromagnetic calorimeters: whereas hadronic calorimeters are built in order to exploit mostly the strong interactions experienced by hadrons (particles containing quarks, such as protons and neutrons) traversing matter, electromagnetic calorimeters detect light particles (electrons and photons) through their electromagnetic interactions with the medium's constituents. Unlike hadronic showers, which are the result of a number of complex hadronic and nuclear processes, the physics of the electromagnetic showers is quite well-understood since it is based on few elementary processes, depending on the nature and energy of the incident particles. More precisely, bremsstrahlung and electron pair production are the dominant processes for high-energy electrons and photons: above 100 MeV electrons and positrons radiates photons (process called bremsstrahlung) as a result of the interaction with the nuclear Coulomb field; on the other hand, in the same energy range, photon interactions produce mainly electron-positron pairs. As a consequence, electrons and photons of sufficient high energy incident on a block of material create secondary photons and electronpositron pairs, which may in turn produce other particles through the same mechanisms. The result is a shower that may consist of thousands of different particles with progressively degraded energies. A diagram of an electromagnetic shower initiated by an electron is shown schematically in Fig 1.

This multiplication process is arrested when the energy of the secondary electrons produced in the electromagnetic cascade falls below a critical energy ε, which may be defined as the energy at which the average energy losses from bremsstrahlung equal those from ionization. At this energy the electrons and positrons lose their energy through collisions with atoms and molecules of the absorber medium, causing ionizations and thermal excitation, while photons are more likely to lose their energy through Compton and photoelectric interactions. When the critical energy is reached, the shower contains the maximum number of particles; the depth at which this occurs is called shower maximum. Since calorimeters have to measure the energy lost by particles that go through them, they are usually designed to entirely stop or absorb the incident particles, forcing them to deposit most of their energy within the detector. Depending on the particular construction technique, the energy lost by the incident particles is collected in the form of light or charge, producing a physical signal proportional to the amount of energy deposited.

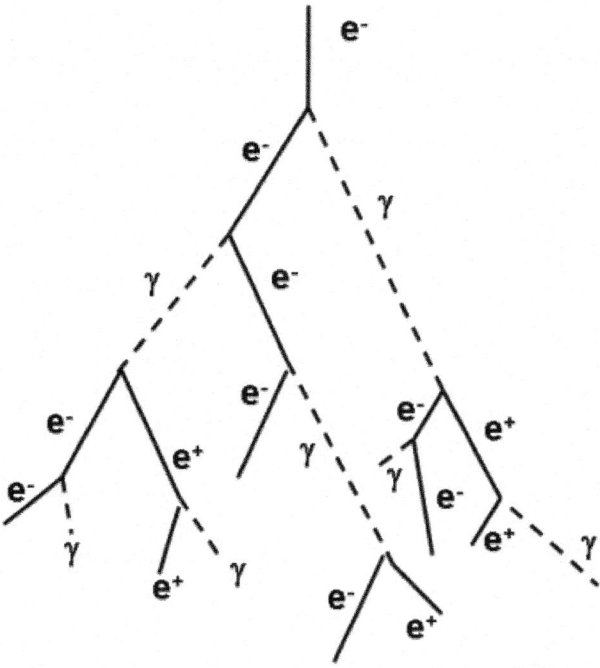

Figure 1: Schematic diagram of an electron initiated electromagnetic shower.

All the energy loss mechanisms that characterize an electromagnetic shower can be calculated with a high degree of accuracy through Quantum Electrodynamics (QED) calculations. For this reason the main features of electromagnetic showers are well known and both the longitudinal and lateral profiles of the showers may be parameterized with simple empirical functions in terms of two parameters, the radiation length X_0 and the Molière radius ρM. The first is defined as the mean distance over which a high-energy electron loses all but $1/e$ of its energy by bremsstrahlung or, in case of photons, as $7/9$ of the mean free path for pair production. The radiation length only depends on the characteristics of the traversed material and it is used to describe the longitudinal development of the shower in a material-independent way. On the other hand the second parameter, the Molière radius, is a measurement of the transverse size of a shower. More precisely it is the average lateral deflection of electrons at the critical energy after traversing one radiation length. The theoretical background needed for understanding the principle which electromagnetic calorimeters are based on is of great help for the evaluation of the performance characteristics of a real electromagnetic calorimeter, especially during the designing phase. The size, the construction principles and the materials used for an electromagnetic calorimeter may be accurately chosen according to the environment in which it has to operate and the tasks it has to fulfil. Calorimetry is the art of compromising between conflicting requirements, such as the energy and spatial resolution, the triggering capabilities, the radiation hardness of the materials used, the dynamic range and so on. Some more details about these aspects will be given in the following sections, together with some practical details of building and operating these detectors.

Homogeneous and Sampling Calorimeters

From the construction point of view, the possible solutions for calorimeters is very wide and quite ingenious calorimeter systems have been designed to cope with more and more demanding physics goals and requirements. Here we will divide calorimeters only in two broad categories, sampling and homogeneous calorimeters.

Sampling Calorimeters

Sampling calorimeters consist of layers of passive or absorber high–density material (lead for instance) interleaved with layers of active medium such as solid lead-glass or liquid argon. Absorber layers are used to enhance photon conversions, while active layers to sample energy loss.

The main drawback of these devices is their limited energy resolution which rises from the large fluctuations, caused by the absorbers, in the energy deposited in the active layers. On the other hand, their excellent space resolution (i.e. their capability to reconstruct the impact position of incident particles) and their satisfactory particle identification are the result of the laterally and longitudinally segmentation, that is relatively easy to implement. Many types of sampling calorimeter exist, which differ one another in the type of materials used. The most common absorber materials are lead, copper and iron, while the active medium can be solid (scintillator and semiconductor), liquid or gaseous. In sampling calorimeters the energy deposited by showering particles can be collected both in the form of light, as in case of scintillation calorimeters, and in the form of electric charge, as happens in gas, solid-state and liquid calorimeters. Some details about the techniques used to collect the light signal are given in Section 3.1.

Homogeneous Calorimeters

In homogeneous calorimeters the same medium is used both to cause the shower development and to detect the produced particles. As discussed later, the main advantage of these devices is their excellent energy resolution, since the intrinsic fluctuations that occur in the development of the showers are small with respect to sampling calorimeters. Usually homogeneous calorimeters are difficult to be segmented and this reduces their capabilities to give information about the position of the incident particle and its identification.

Since the materials used to build homogeneous calorimeters are characterized by large interaction lengths, such devices are almost exclusively used for electromagnetic calorimetry. Homogeneous calorimeters can be classified, according to the type of active material, into semiconductor calorimeters, noble-liquid calorimeters, Cherenkov calorimeters and scintillator calorimeters. Whereas the first two types of devices are based on the charge measurement, in scintillator and Cherenkov calorimeters the signal is collected in the form of light: the photons produced are converted into electrons and the electric signal is amplified to reduce the electronic noise level. This is usually performed by photo-sensitive devices, such as photomultipliers, which are able to reach multiplication gains of the order of 106. However, the use of magnetic fields in modern particle experiments prevents the use of traditional photomultipliers: in these cases avalanche photodiodes are a valid alternative; however they provide a moderate gain, causing a non-negligible noise term.

The choice of the typology and of the detector parameters for a specific application depends on several factors, such as the physics goals, the energy range that has to be considered, the accelerator characteristics and the available budget as well. Often the choice is supported by results from tests of small prototypes with particles beams, as well as by detailed simulation studies of the calorimeter performance.

The Energy Response of a Calorimeter

In an electromagnetic calorimeter, the energy dissipated by the charged particles of a shower in the detector material is converted into a detectable signal which provides information on the original energy of the incident particle. This is verified both in case of homogeneous and sampling calorimeters: in the former, the whole energy of an incident particle is deposited in the active material, so that the entire detection volume may contribute to the signals that the particle generates; in the latter, the presence of passive absorber layers reduces the energy lost in the active layers to a fixed fraction (called sampling fraction) of the original energy.

However, the energy response of calorimeters depends on the nature of the incident particle: whereas showering electrons and photons produce a signal that is proportional to the original energy, the same is most certainly not valid for hadrons because part of their energy is used to dissociate the atomic nuclei and does not contribute to the calorimeter signals. Hence, the hadronic signals from electromagnetic calorimeters are non-linear and are not constant as a function of energy. In the following we will focus exclusively on the response of electromagnetic calorimeters to electron and photons.

In order to provide a reliable measurement of the energy, a first requirement that has to be fulfilled concerns the containment of the shower inside the detector volume. As the energy of the incident particle increases, the detector size needed to contain the showers increases as well. For electromagnetic showers, a simple formula that gives approximately the location of the shower depth expressed in radiation lengths is the following:

$$x_{max} \approx \ln(E/\varepsilon) + x_0 \tag{1}$$

where E is the energy of the incident particle, and $x_0 = 0.5$ for photons and -0.5 for electrons. This expression shows that the longitudinal size of a shower increases only logarithmically with energy, allowing to design a compact detector even for electromagnetic showers depositing a large energy (hundred of GeV) into the sensitive volume. On the other hand, the thickness containing 95% of the total shower energy is approximately located at $x_{95\%} \approx x_{max} + 0.08Z + 9.6$, indicating that the realization of compact calorimeters requires the use of high-Z materials.

Therefore, even at the particle energies reached at the Large Hadron Collider, electromagnetic calorimeters are very compact devices and energetic showers lose only a small percentage of their energy beyond the end of the active calorimeter volume. Besides being intrinsically linear, electromagnetic calorimeters should also give a precise measurement of the energy. The precision with which the unknown energy of a give particle is measured is called energy resolution and represents one of the most important performance characteristics of a calorimeter. The actual energy resolution of a realistic electromagnetic calorimeter can be in general parameterized as follows:

$$\sigma/E = a/\sqrt{E} \oplus b/E \oplus c \tag{2}$$

where σ is the standard deviation in the energy measurement, the constants a, b and c depend on the detector characteristics and the symbol $\oplus$ indicates a quadratic sum. From the left-hand side, the three terms are known as stochastic term, noise term and constant term respectively; the importance of each term strongly depends on the energy deposited in the calorimeter.

The stochastic term arises from the shower intrinsic fluctuations that characterize electromagnetic cascade developments event by event. Unlike homogeneous calorimeters, where these fluctuations are moderate, the stochastic term is quite important especially in sampling calorimeters, due to the presence of the absorber layers. The noise term arises from the electronic noise of the readout chain. Usually it does not weight much upon the total energy resolution, especially for calorimeters in which the signal is collected in the form of light because, in that case, the first step of the electronic chain is a photo-sensor (like a phototube or a Silicon photomultipliers) which amplifies the original signal with almost no noise.

However, for energetic particles as those produced in the new-generation accelerators, the dominant contribution to the energy resolution is the constant term, that does not depend on the energy of the particle and arises from systematic effects (such as detector nonuniformities, shower leakage, mis-calibration,…). Therefore, modern calorimeters are built imposing severe construction tolerances in order to reduce possible instrumental imperfections that may give rise to response non-uniformities.

Additional contributions can make the energy resolution of a calorimeter worse, such as lateral and longitudinal leakages of the energy shower outside the active calorimeter volume, or fluctuations due to energy losses of electrons and photons in inactive materials (mechanical structures and cables) before reaching the calorimeter. These effects mainly affect the energy resolution of calorimeters integrated in complex high-energy physics experiments where the calorimeter is only one component that has to satisfy several mechanical constraints.

The tasks of an em-calorimeter

Calorimeters were originally conceived as devices used to measure only the energy of the incident particles. Today they have different applications and often their tasks are made more effective when the information they provide is combined with that coming from other sub-detectors.

The reasons that make calorimeters so attractive are various. First of all they provide a precise measurement of the energy in a wide range, since the energy resolution varies with energy as $1/\sqrt{E}$. This characteristic is crucial when looking for narrow resonances characterized by a poor signal/background ratio, as in the case of the search for Higgs bosons decaying into $\gamma\gamma$. Moreover, the capability to entirely absorb high energetic showers in compact distances make electromagnetic calorimeters the suitable instrument to measure electrons and photons over an unprecedented energy range, as requested by the experiments running at recent accelerator facilities like the LHC.

Electromagnetic calorimeters can also provide a fairly good particle identification, being able to distinguish, for instance, electrons and photons from muons and pions on the basis of their different energy deposit profiles. Moreover, calorimeters are sensitive to all kind of particles, including neutral ones. This feature is particularly exploited in particle physics experiments for several applications: for example, when measuring jets of particles (i.e. the characteristic collimated sprays of hadrons coming from hard collisions between energetic partons), electromagnetic calorimeters can give a measurement of the neutral component of the jets, completing the information provided by other kinds of detectors sensitive only to charged particles. Moreover, in modern experiments, calorimeters are built in order to have an angular coverage as large as possible, achieving often hermeticities in excess of 90%; in this way they can provide an indirect measurement of weakly interacting particles, as neutrinos, that can be detected only by observing geometric imbalances (missing energy) in the total transverse energy.

The possibility to segment electromagnetic calorimeters into several identical cells allows to gain information also on the impact position of the incident particles. The shower position is determined by reconstructing the centre of gravity of the energy deposited in the various detector cells that contribute to the signal. The precision with which the position of a given particle is measured mainly depends on the granularity of the detector and can achieve accurate values. The granularity of an electromagnetic calorimeter plays also an important role for the minimization of the pileup phenomena, i.e. the overlap of signals coming from different particles hitting the same cell. This problem may occur in calorimeters running in high-density particle environments characteristic of ultra-relativistic heavy-ion collisions. Moreover, the high luminosities achieved in modern hadron colliders may cause another type of pileup due to the overlap of signals coming from the preceding or following bunch crossing. To avoid event overlaps the calorimeter response in time is usually fast enough. This important feature is extensively used to select or trigger on interesting events, especially in those experiments in which the event rate is orders of magnitude beyond the rate at which events may be collected by the data acquisition system. Signals from calorimeters are available at very short time after the particle impact and are easy to process and interpret: the characteristics derived from the calorimeter data allows physicists to select only interesting (and often rare) events, such as those containing jets of particles or very energetic electrons and photons.

Thanks to all these features, calorimeters became key components of particle detectors and today almost every experiment in particle physics relies heavily on calorimetry.

READOUT SYSTEMS BASED ON AVALANCHE PHOTO-DIODES: PROBLEMS AND SOLUTIONS

Light collection in Electromagnetic Calorimeters

In order to collect all the energy deposited in each calorimeter cell, a proper readout system must be designed, to convert the light produced in the active material into an electronic signal which can be handled by the analog-to-digital converters and/or used by the trigger logic. Any readout system must be designed in such a way not to disturb the working condition of the calorimeter. For instance, no additional effect has to be introduced as far as the energy, position and time resolution are concerned. Moreover, the hermeticity of the calorimeter needs not be degraded by the readout part of the detector, as well as the overall radiation hardness.

In case of homogeneous calorimeters, which usually consist of individual crystals, scintillation or Cerenkov light is generated by the passage of particles. Each crystal is read out from the back, by the use of a photomultiplier tube (PMT) or by a different light sensor. While traditional readout devices employed photomultipliers in the past, solid-state devices started to be used more than 15 years ago. Silicon photodiodes (Barlow et al., 1999), Hybrid Photon Detectors (Anzivino et al., 1995) or APDs (Lorenz et al., 1994) are among the oldest examples of such devices.

Sampling calorimeters have been used since the very beginning in particle physics. In these structures, metal plates are interleaved with active scintillation materials. The geometry of the first generation calorimeters was very simple, with the scintillation light being extracted by each scintillation portion of the detector, coupled to one or two photomultipliers. In recent years however, the geometry of sampling calorimeters has changed considerably, mainly to design calorimeters with hermeticity properties. The development of wavelength shifting WLS fibers has allowed to design calorimeters with a fine segmentation of the active material, without enlarging the number of readout channels. In sampling calorimeters, the WLS fibers run along the length of each cell or perpendicular to it, and all the portions of the scintillation material are optically coupled to a photon detector located behind the calorimeter. This has the advantage that all the active parts of a cell are read out by the same photo-sensor. Some disadvantages inherent to this technique are related however to the worsening of the signal timing, which is much slower than that of the scintillation process, and to the partial non-hermeticity introduced by the volume itself of the WLS fibers. The latter may be solved with the use of scintillating and wavelength shifting fibers.

As far as the time structure of the signal is concerned, the absorption of the electromagnetic shower induced in a calorimeter by a high energy particle or radiation takes a time in the order of a few nanoseconds, due to the typical size of each individual module. However, the detection of neutrons has a much longer time scale, since they may be scattered from the surrounding materials. As a consequence, a long tail may be observed, which also requires some intervention on the associated electronics, such as to modify the shaping time of the signal in order to cut a large fraction of this tail.

In addition to the late arrival of neutrons, other instrumental effects may also have their influence on the time structure of the signals. One of these effects is due to the nature of the particular scintillation material used as active element in the calorimeter, since the fluorescent processes by which light is produced inside a scintillation material have characteristic decay times which range from nanoseconds to milliseconds. While organic scintillators have decay times of a few nanoseconds, considerably longer times (300 ns) are typical of BGO material.

An important aspect to be considered in understanding the time structure of the signal – especially for large volume calorimeters - is the fact that relativistic particles which produce the light inside the scintillation detector cover the distance between the front face of the detector and the photo-detector in a time which is smaller than the time required to the light photons produced by their passage, due to the refraction index n of the material. Such difference between c and c/n may amount to a few nanoseconds, which may be relevant in case of fast scintillators.

Avalanche Photodiodes as photo-sensors in calorimeters

Silicon Avalanche Photodiodes are now considered good candidates to replace traditional photomultipliers to detect the light produced in scintillation materials, especially in specific situations, where their advantages become more apparent. APDs have several features which make them attractive for scintillation detection. The discussion of the working principles of such devices and their detailed structure is beyond the scope of the present Chapter. Here only follows a list of different aspects which are relevant for the use of APDs as photo-sensors in electromagnetic calorimeters, and which need to be evaluated when choosing the proper solution:

Size: Photomultipliers have different sizes, which can be adapted in principle to any specific situation. However, the size of traditional photomultipliers is always much bigger than their sensitive area, which can be a problem for large detector arrays, as it is the case with large particle detectors. On the contrary, APDs have limited sizes (in the order of a 1-50 mm2) and the required front-end electronics may also be very compact. For such reasons, their use is suitable to the need of a large granularity crystal array (in case of homogeneous or sampling calorimeters). On the other side, their small size may require sometimes to employ more than one device per crystal, in order not to reduce too much the amount of collected light.

Internal gain

APDs are components which possess an intrinsic internal gain. The applied bias voltage produces a region where a high electric field (in the order of 150 kV/cm) is obtained. Such field is able to generate avalanches of secondary particles, and hence to amplify the signal. The exact value of the internal gain coefficient depends on the applied bias voltage and also on the temperature. In comparison to the possibility of "external" gain, the intrinsic gain has the advantage of using all the signal generated by the light development inside the scintillation crystal, whereas the signals generated within the sensitive volume of the diode are not amplified.

Insensitivity to magnetic field

Due to the need to bend charged particles emitted in a nuclear collision and hence measure their momentum according to the curvature of the track, a large fraction of a modern particle detector is usually contained inside a magnetic field. Large magnetic fields, up to several Tesla, may be in order, with the aim to bend very energetic (GeV or more) particles. Such fields however do not allow the use of traditional photomultipliers, since electron trajectories would be highly distorted by the field, and even robust shielding could be ineffective to protect them. Moreover, the use of a thick shielding would cause a large number of secondary interactions in the material, which could be of disturb for the detection of particles and radiations in the active detectors. The use of traditional PMTs would then require to transport the light through optical fibers for a long distance outside the magnetic field. APDs on their side are not sensitive to magnetic fields, which makes them the device of choice for particle detectors embedded in large magnets.

Spectral response

While the spectral response of an Avalanche Photodiode is nearly the same as for a normal photodiode when no bias is applied, this changes as a result of the application of the bias voltage, since the penetration depth of the light inside the device depends on the wavelength. Devices which have an enhanced sensitivity in the near-infrared region or at smaller wavelengths (300 nm) are available nowadays, to allow the user to select the appropriate device.

Quantum efficiency

The overall efficiency of a photo-sensor strongly depends on how good is the matching between the emission spectrum of the scintillation material being used and the spectral response of the photosensor. Avalanche photodiodes usually exhibit a very high quantum efficiency (QE), relatively constant over a wide wavelength interval. As an example, the APDs employed in the ALICE and CMS calorimeters exhibit a quantum efficiency of the order of 85% from 500 to 800 nm, dropping down to 50% at 350 and 950 nm. These values are significantly higher than the average values for standard photomultipliers, which are in the order of 20%.

Gain stability

The gain of an Avalanche Photodiode depends both on the applied voltage and on the temperature, so it is important to measure, for any individual APD, the relevant coefficients, in order to allow for possible variations in these quantities. The stability of the bias voltage depends on the overall quality of the power supply and (to some extent) on the environmental conditions. While it is not so critical to control the bias voltage, a relevant problem is that the gain of each APD needs to be individually adjusted to match a common value of the gain, which requires a software-controlled power supply, able to distribute to each APD channel the appropriate voltage. Concerning the temperature dependence, this is an important factor to maintain the gain stable over time, since the ambient temperature may be subjected to non negligible variations over long time operational periods. Moreover, due to the large number of channels usually involved, sensible temperature differences are experienced by devices which are located even meters apart, so that a monitoring and correction of the bias voltage to overcome such temperature-dependent gain variations is mandatory to equalize the gain in different channels.

Measurements have been also performed (Chartrchyan et al. 2008) on the long term stability of such devices, maintaining them under bias voltage for long periods, up to 250 days, and checking for possible failures through the monitoring of the dark current. Values of the MTTF (mean time to failure) of the order of 107 hours have been reported.

Negligible nuclear counter effect

The nuclear counter effect is related to the amount of extra signal produced inside the photodiode by a charged particle traversing it, which adds to the charge produced by the scintillation light in the crystal. It can be quantified by the thickness of a Si PIN diode required to produce the same signal. The signal produced in the APD would be proportional to such equivalent thickness. From this point of view, the effective thickness of the device should be minimized as much as possible in order to reduce the influence of the nuclear counter effect. However, reducing the thickness has also the effect of increasing the APD capacitance, so that a compromise needs to be reached.

Small excess noise

Avalanche photodiodes generate excess noise, due to the statistical nature of the avalanche process. An estimate of the statistical fluctuations of the APD gain is given by the excess noise factor F, where $\sqrt{F}$ is the factor by which the statistical noise on the APD current exceeds that expected from a noiseless multiplier. The excess noise has its origin in the intrinsic statistical nature of the internal charge carrier multiplication inside the device, which depends on the inhomogeneities in the avalanche region and in hole multiplication. If k is the ratio of the ionization coefficients for electrons to holes, at a given gain M, the excess noise factor is given by:

$$F = k \times M + (2 - 1/M) \times (1-k) \tag{3}$$

The result is an additional contribution to the energy resolution, and clearly a small value of the excess noise factor is preferable to optimize the overall resolution. This factor increases with the gain, reaching for instance a value of about 1.9 at M=30 for the APD employed in the ALICE and CMS calorimeters. Large area APDs which have been subsequently developed for the PANDA calorimeter, exhibit smaller values of F (1.38 at M=50).

High resistance to radiations

The use of Avalanche Photodiodes in hostile environments, as far as the radiation level is concerned, is a critical point for large particle physics experiments, where the flux of charged and neutral particles produced in high energy collisions over long operational periods may be very high. The dose absorbed by the detectors and associated electronics is usually evaluated by detailed GEANT simulations which take into account the description of the complex geometry and materials of the detector. Depending on the physics program (proton-proton or heavy-ion collisions, low or high beam luminosity, allocated beam time,...) and on the location of such devices inside the detector, a particular care must be devised to understand whether the photo-sensors will be able to survive during the envisaged period of operation. For such reason, a detailed R&D program has been undertaken within the High Energy Collaborations to expose the devices of interest to different sources of radiations, and measure their performance before and after irradiations. There are basically two damage mechanisms: a bulk damage, due to the displacement of lattice atoms, and a surface damage, related to the creation of defects in the surface layer. The amount of damage depends on the absorbed dose and neutron fluence.

Whereas experiments like ALICE, which will run with low luminosity proton and heavy ion beams at LHC, do not suffer of big problems with the radiation dose in the electromagnetic calorimeter, the CMS detector, which runs at a much larger luminosity, will have a very large dose in the photo-sensors. As an example, in ten years LHC operation, the planned dose in the CMS barrel is in the order of 300 Gy, with a neutron fluence of 2×10^{13} n/cm^2 (1 MeV-equivalent). This has lead to an extensive set of measurements with different probes (protons, photons and neutrons), an to the successful development of APDs capable to survive to these conditions.

Front-end electronics

Once the light produced in the active material has been collected by the photosensor, an important step towards the extraction of the signal is the associated front-end electronics. Such electronics has to be used to process the signal charge delivered by the photo-sensors and extract as much information as possible concerning the time and amplitude of the signal. Several aspects are important to understand the requirements which are demanded to front-end electronics.

Dynamic range

In high energy experiments, for instance in the experiments running at LHC, the dynamic range required to a calorimeter is very high. Signals of interest go from the very small amplitudes associated to MIP particles (for instance, cosmic muons used for the calibration, which typically deposit an energy of a few hundred MeV in an individual cell) to highly energetic showers (in the TeV region) produced by hadrons or jets. The dynamic range required may then easily cover 4 orders of magnitude, which requires a corresponding resolution in the digitization electronics (ADC with 15-16 bits). An alternative approach is the use of two separate high-gain and low-gain channels, which requires ADCs with a smaller number of bits, at the expense of doubling the number of channels.

Time Information

The extraction of timing information from the individual signals originating from each module in a segmented calorimeter is an important goal for the front-end electronics. Time information may be important in itself, also for calibration and monitoring purposes, and it is mandatory when the information from a calorimeter must be used to provide trigger decisions. The timing performance of the overall readout system also depends on the rest of the electronics, as well as on the algorithms being used to extract such information (See Sect.6).

Number of independent channels

Due to the large granularity usually employed in segmented calorimeters, the number of independent channels is very high, in the order 104-105. This requirement demands a corresponding high number of front-end preamplifiers and a high level of integration for the associated electronics, which needs to be compacted in a reasonable space.

Monitoring systems

A common aspect to all kind of detectors which are used to transform the light, produced in the active part of the calorimeter, into an electric signal, is the fact that their exact response (gain) is intrinsically unstable, depending on a number of factors which may vary according to the experimental conditions. Temperature and voltage variations are particularly important in this respect, as discussed before, since the gain of Avalanche Photodiodes is very sensitive to such parameters. Such aspects require usually a careful study of the devices being used, under the specific working conditions, in order to characterize their response as a function of these parameters (see Sect.5). Moreover, a monitoring system is in order, to take into account the variation of the working parameters, and sometimes even to correct the gain by a proper feedback. A LED monitoring system is usually employed in large calorimeters, with the aim to send periodically a reference signal to all readout cells and to check the response uniformity.

A REVIEW OF LARGE APD-BASED ELECTROMAGNETIC CALORIMETERS

Most of the large experiments devoted to high energy physics make use of calorimeters, to detect hadronic and electromagnetic showers originating from energetic particles and radiations. Electromagnetic calorimeters in particular are employed since several decades, making use in the past of traditional photo-sensors (photomultipliers) and, more recently, of solid-state devices such as photodiodes, APD and silicon photomultipliers. Here a brief review is given of several experiments in high-energy physics which have an electromagnetic calorimeter as an important part of the detection setup.

Calorimeters based on traditional photo-sensors

Several high-energy experiments installed in the largest nuclear and particle physics Laboratories have employed in the past electromagnetic calorimeters of various configurations and design, with traditional photomultipliers or photodiodes as photon sensitive devices. As an example, Table 1 shows a (non-exhaustive) list of detectors which include an electromagnetic calorimeter, together with some basic information on the organization and design of the detector. As it can be seen, the largest installations have a number of channels in the order of 10^4, which is remarkable for traditional readout systems based on photomultipliers.

Experiment	Laboratory	Type	No.of channels
E731	FNAL	Lead Glass	802
CDF	FNAL	Lead/Scint	956
FOCUS	FNAL	Lead/Scint	1136
SELEX (E781)	FNAL	Lead Glass	1672
BABAR	SLAC	CsI (photodiode)	6580
L3	CERN /LEP	BGO Crystals (photodiode)	10734
OPAL	CERN /LEP	Lead Glass	9440
HERMES	DESY /HERA	Lead Glass	840
HERA-B	DESY/HERA	Pb(W-Ni-Fe)/Scint Shashlik-type	2352
H1	DESY/HERA	Lead-scintillating fibre	1192
ZEUS	DESY/HERA	Depleted uranium-Scint calorimeter, WLS	13500
WA98	CERN /SPS	Lead Glass	10080
KLOE	LNF	Lead-scintillating fibre	4880
STAR	RHIC	Pb/Scint Sampling calorimeter, WLS	5520
PHENIX	RHIC	Pb/scint shashlik-type	15552
PHENIX	RHIC	Pb glass	9216
LHCb	CERN /LHC	Lead/Scint shashlik-type, WLS	5952

Table 1: Summary of detector installations which make use of an electromagnetic calorimeter with traditional readout devices

Calorimeters making use of Avalanche Photodiodes

Only in the last years Avalanche Photodiodes have been routinely employed as photosensors for large electromagnetic calorimeter installations. Here we want to briefly summarize a few examples of recent detectors which have been installed and commissioned or in the stage of being constructed.

The electromagnetic calorimeter of the CMS experiment at LHC

CMS (Compact Muon Solenoid) is one of the large experiments running at the CERN Large Hadron Collider (LHC). A general description of the CMS detector is reported in (Chartrchyan et al. 2008). A large electromagnetic calorimeter, based on lead tungstate crystals with APD readout, is included in the design of the CMS detector. The barrel part of the CMS electromagnetic calorimeter covers roughly the pseudo-rapidity range $-1.5 < \eta < 1.5$, with a granularity of 360-fold in φ and 2x85-fold in η, resulting in a number of crystals of 61200. Additional end-caps calorimeters cover the forward pseudorapidity range, up to $\eta=3$, and are segmented into 4 x 3662 crystals, which however employ phototriodes as sensitive devices.

The use of lead tungstate crystals with its inherent low light yield and the high level of ionizing radiations at the back of the crystals has precluded in this case to employ conventional silicon PIN photodiodes. In collaboration with Hamamatsu Photonics, an intensive R&D work has led the CMS Collaboration to the development of Si APDs particularly suited to such application (Musienko, 2002). As a result of this work, a compact device (5x5 mm^2 sensitive area, 2 mm overall thickness) has been produced, which is now used also by other experiments. The performances of such device are its fast rise time (about 2 ns) and the high quantum efficiency (70-80 %), at a reasonable cost for large quantities. To overcome the inherent limitations of a reduced gain at wavelength smaller than 500 nm, and a high sensitivity to ionizing radiation, an inverse structure for such devices was implemented. In these APDs the light enters through the p^{++} layer and is absorbed in the p^+ layer. The electrons generated in such layer via the electron-hole generation mechanism drift toward the pn junction, amplified and then drift to the n^{++} electrode, which collects the charge. The APD gain is largest for the wavelengths which are completely absorbed in the p^+ layer, which is only a few micron thick; as a result, the gain starts to drop above 550 nm. Moreover, with this reverse structure, the response to ionizing radiation is much smaller than a standard PIN photodiode.

An important issue for the APD installed in the CMS detector is the effect of radiation on the working properties of the device, due to high luminosity at which this experiment is expected to run for most of its operational time. In ten years of LHC running, the neutron fluence (1 MeV equivalent) in the barrel region is expected in the order of 1013 n/cm², with a dose of about 300 Gy. The extensive irradiation tests performed in the context of this Collaboration have provided evidence that the devices are able to survive the long operational period envisaged at LHC.

Due to the large area of the crystals employed in the CMS calorimeter, compared with the sensitive area of the APD devices, two individual Avalanche Photodiodes are used to detect the scintillation light from each crystal.

The electromagnetic calorimeter of the ALICE experiment at LHC

The ALICE detector (Aamodt et al., 2008) is another large installation at LHC, mainly devoted to the heavy ion physics program. It is equipped with electromagnetic calorimeters of two different types: the PHOS (PHOton Spectrometer), a lead tungstate photon spectrometer, and the EMCAL, a sampling lead-scintillator calorimeter. These two detectors are able to measure electromagnetic showers in a wide kinematic range, as well as to allow reconstruction of neutral mesons decaying into photons.

The PHOS spectrometer is a high resolution electromagnetic calorimeter covering a limited acceptance domain in the central rapidity region. It is divided into 5 modules, for a total number of 17920 individual Lead tungstate (PWO) crystals. Each PHOS module is segmented into 56 x 64=3584 detection cells, each of size 22 x 22 x 180 mm, coupled to a 5 x 5 mm² APD.

An additional electromagnetic calorimeter (EMCal) was added to the original design of ALICE, to improve jet and high-pt particle reconstruction. This is based on the shashlik technology, currently employed also in other detectors. The individual detection cell is a 6 x 6 cm2 tower, made by a (77+77) layers sandwich of Pb and scintillator, with longitudinal wavelength shifting fiber light collection. The total number of towers is 12288 for the 10 super-modules originally planned (which cover an azimuth range of 110°). Recently a new addition of similar modules started, to enlarge the electromagnetic calorimeter (DCAL), providing back-to-back coverage for di-jet measurements. This will roughly double the number of channels.

The active readout element of the PHOS and EMCal detectors are radiation-hard 5 x 5 mm2 active area Avalanche Photodiodes of the same type as employed in the CMS electromagnetic calorimeter. These devices are currently operated at a nominal gain of M=30, with a different shaping time in the associated charge-sensitive preamplifier.

The electromagnetic calorimeter of the PANDA experiment at FAIR

PANDA is a new generation hadron physics detector (Erni et al., 2008), to be operated at the future Facility for Antiproton and Ion Research (FAIR). High precision electromagnetic calorimetry is required as an important part of the detection setup, over a large energy region, spanning from a few MeV to several GeV. Lead-tungstate has been chosen as active material, due to the good energy resolution, fast response and high density. To reach an energy threshold as low as possible, the light yield from such crystals was maximized improving the crystal specifications, operating them at -25 °C and employing large area photo-sensors. The largest part of such detector is the barrel calorimeter, with its 11360 crystals (200 mm length). End-cap calorimeters will have 592 modules in the backward direction and 3600 modules in the forward direction. The crystal calorimeter is complemented by an additional shashlyk-type sampling calorimeter in the forward spectrometer, with 1404 modules of 55 x 55 mm2 size.

The low energy threshold required of a few MeV and the employed magnetic field of 2 T precludes the use of standard photomultipliers. At the same time, PIN photodiodes would suffer from a too high signal, due to ionization processes in the device caused by traversing charged particles. In order to maximize the light signal, new prototypes of large area (10×10 mm^2 or 14×6.8 mm^2), APDs were studied, devoting particular care to the radiation tolerance of these devices. In the forward and backward end-caps, due to the high expected rate and other requirements, vacuum phototriodes (VPT) were the choice. Such devices, which have one dynode, exhibit only weak field dependence, and have high rate capabilities, absence of nuclear counter effect and radiation hardness.

CHARACTERIZATION OF AVALANCHE PHOTODIODES FOR LARGE DETECTORS: PROCEDURES AND RESULTS

As discussed in the previous Sections, the construction of a large electromagnetic calorimeter based on Avalanche Photodiodes as readout devices may require a large number (in the order of 10^3-10^5) of individual APDs to be tested and characterized, after the R&D phase has successfully contributed to produce a device compliant with the specifications required by the experiment. Not only the devices have to be checked for their possible malfunctioning, but to minimize the energy resolution for high energy electromagnetic showers, it is important to obtain and assure a relative energy calibration between the different modules into which the calorimeter is segmented. The uncertainty in the inter-module calibration contributes to the constant term in the overall energy resolution, which becomes most significant at high energy. An additional motivation to have a good module-to-module calibration comes from the possibility to implement on-line trigger capabilities, especially for high energy and jet events. In such case, it is mandatory to adjust the individual gains of the various channels within a few percent.

For all such reasons, a massive work is usually required to choose the optimal APD bias for each individual device. Such massive production tests allow also to check the functionality of the device under test and the associated preamplifier, prior to mounting them in the detector. Mass production tests carried out in the lab prior to installation usually consist of measurements of the gain versus voltage dependence of each APD at fixed and controlled temperature, and in the determination of the required voltage to reach a uniform gain for all the devices.

Several properties may be measured during this screening operation, depending on the amount of information required, the desired precision and the amount of time at disposal to carry out all the required operations in a reasonable time schedule. If the device under consideration originates from a stable production chain at the manufacturer's site, as it is usually for APDs which have been in use for several applications, a complete set of characterization procedures may be carried out only for limited samples of devices. These may include the evaluation of the quantum efficiency, of the excess noise factor, of the capacitance, dark current and gain uniformity over the APD surface, as well as the temperature dependence of the gain curve in a wide range of temperatures (Karar, 1999). Massive tests, to be carried out on each individual APD, at least require the measurement of the gain-bias voltage curve at one or more temperatures, close to the operational one, and (possibly) the measurement of the dark current at different gain values. From the measured data one can extract the bias voltage required to match a fixed value of the gain, and the voltage coefficient.

The basic equipment to carry out such tests includes a system to maintain and measure the APD temperature while performing the measurements (usually within 0.1 °C), a pulsed light source (for instance a pulsed LED in the appropriate wavelength region), the front-end electronics and some acquisition system to store the data for further analysis. Due to the large number of devices usually under test, a suitable procedure must be designed, which tries to minimize as much as possible the time required to carry out a complete scan. As an example, the test of several APDs (8-32) at the same time may be planned with a proper choice of the readout system. Moreover, bias voltage may be software controlled together with acquisition, thus allowing to carry out automatic measurements in controlled steps of bias voltage. Fig.2 shows an example of a typical gain curve obtained during the characterization of a large number of Hamamatsu S8148 APDs within the ALICE Collaboration (Badalà, 2008). The output signal was measured for different values of the bias voltage, from 50 V (where a plateau is expected, corresponding to unitary gain) to about 400 V. The data were fitted by the function:

$$M(V) = p_0 + p_1 \exp(-p_2 V)$$

(4)

in order to extract the coefficients p_0, p_1, p_2 and thus determine the voltage V_{30} at which the gain equals M=30, which is the required value in the ALICE EMCal.

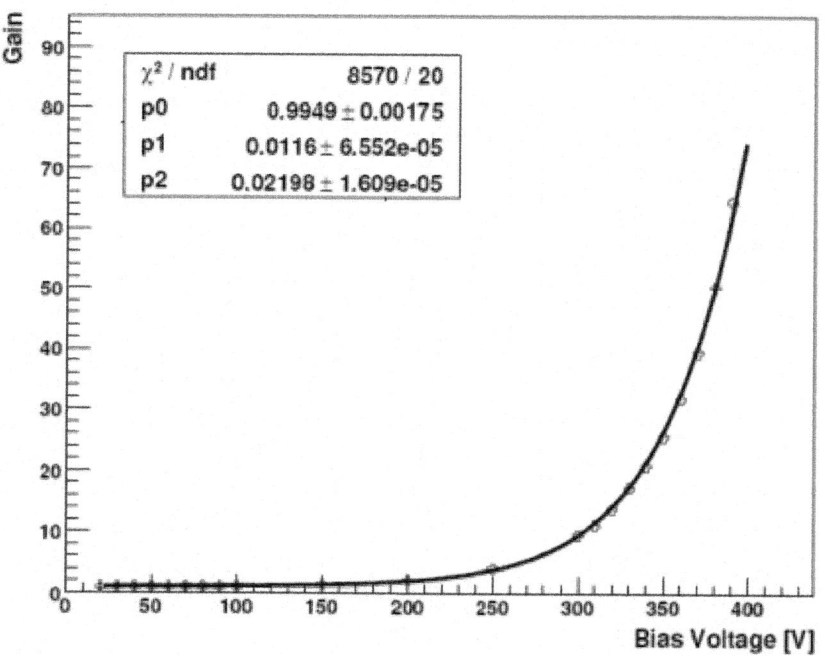

Figure 2: Gain curve as a function of the APD bias voltage, for one of the Hamamatsu S8148 employed in the ALICE electromagnetic calorimeter. A common gain of 30 is usually set for all the modules.

The relative change in the gain with the bias voltage is an important parameter to extract from such measurements, especially in the region where the APD will work. Fig.3 reports one of such results, showing a value of 2.3 %/V at M=30.

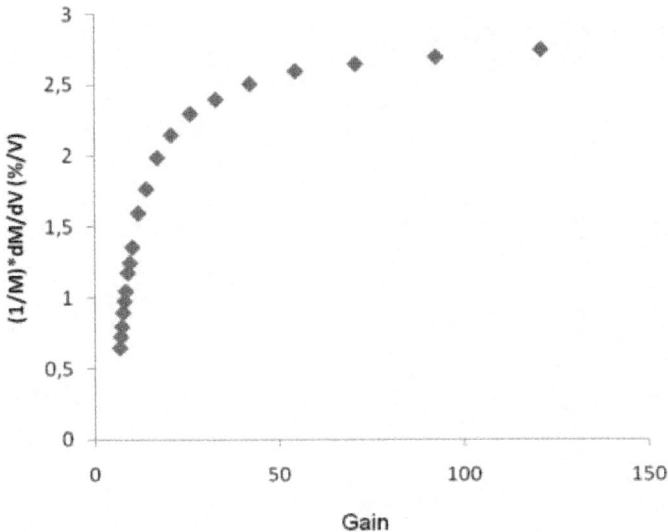

Figure 3: The relative change in the APD gain is here reported at different values of the gain

Due to the strong dependence of the APD gain from the temperature, the investigation of the gain versus temperature is an important issue of the characterization phase, at least for subsamples of the complete set of devices. Gain curves have to be measured for different values of the temperature – spanning the region of interest - in order to extract a temperature coefficient. Fig.4 shows an example of a set of different gain curves measured in the range 21 to 29 °C, for the Hamamatsu S8148 APDs.

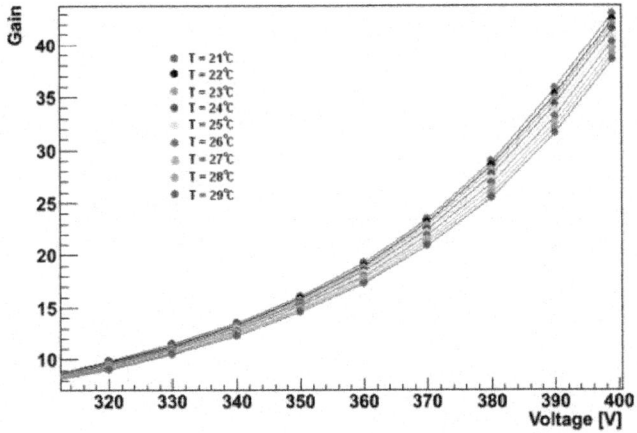

Figure 4: Gain curves measured at different temperatures.

This or similar sets of measurements allow to extract the gain versus temperature dependence (Fig.5) and finally a value of the temperature coefficient, which decreases with the temperature, as shown in Fig.6.

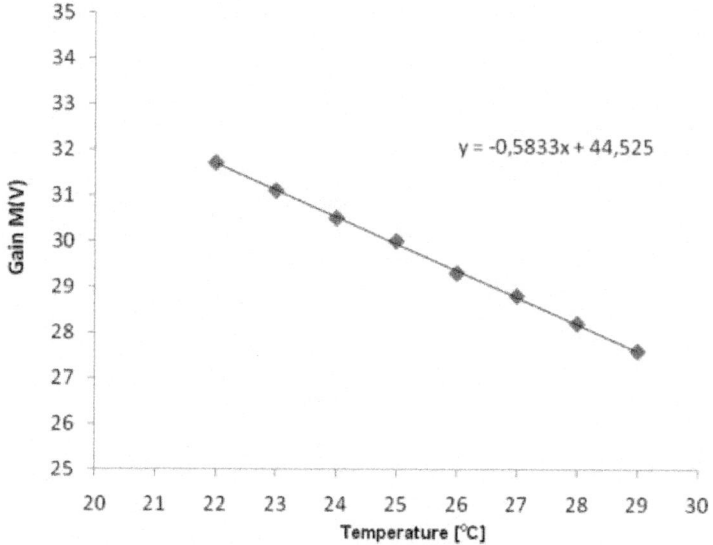

Figure 5: APD gain as a function of the temperature.

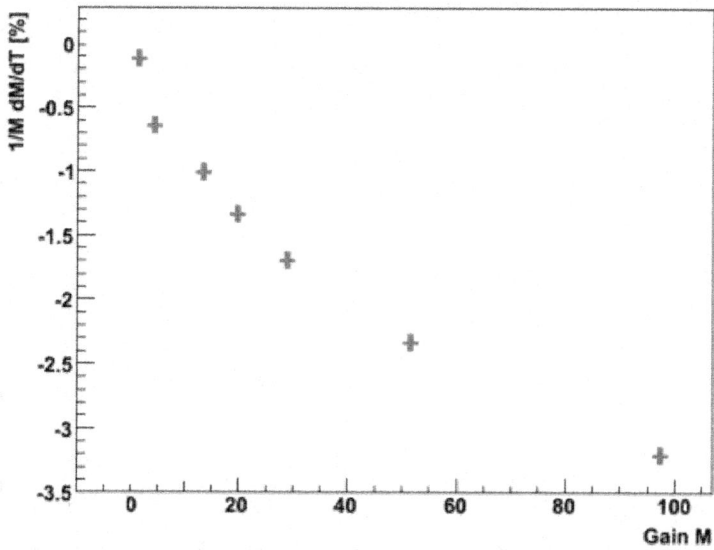

Figure 6: Temperature coefficient of the APD gain, reported as a function of the APD gain.

All these procedures allow to classify the individual devices into different categories (for instance according to the voltage required to match a given gain, or to the temperature coefficient) for the sake of response uniformity, and to reject APDs with inadequate performance. Carrying out systematic characterization of a large number of individual devices permits to investigate statistical distribution of several quantities of interest, and establish classification criteria, to be used for the next samples. As an example, Fig.7 shows the distribution of the bias voltages required to have a common gain (M=30) in a set of 1196 APDs which were used in one of the super-module of the ALICE electromagnetic calorimeter.

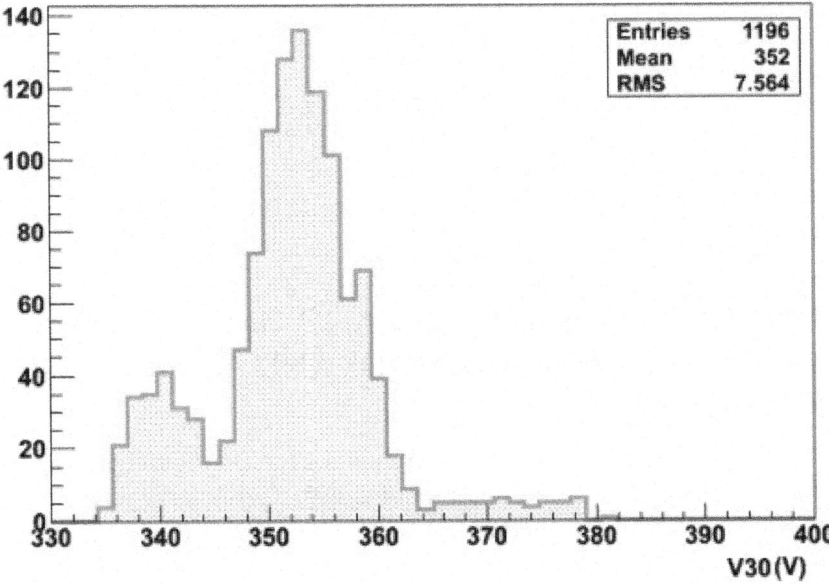

Figure 7: Statistical distribution of the APD bias voltages required to match a common gain M=30, for a set of 1196 devices employed in one of the super-modules of the ALICE calorimeter.

While the distribution shows clearly the presence of two populations (due to different production lots), all devices showed a bias voltage smaller than 400 V, which was the limit set by the electronic circuitry to power the APD with a sufficient resolution. Fig.8 shows also the distribution, for the same set, of the voltage coefficient, which has an average value of 2.3%/V, with an RMS in the order of 0.08 %/V.

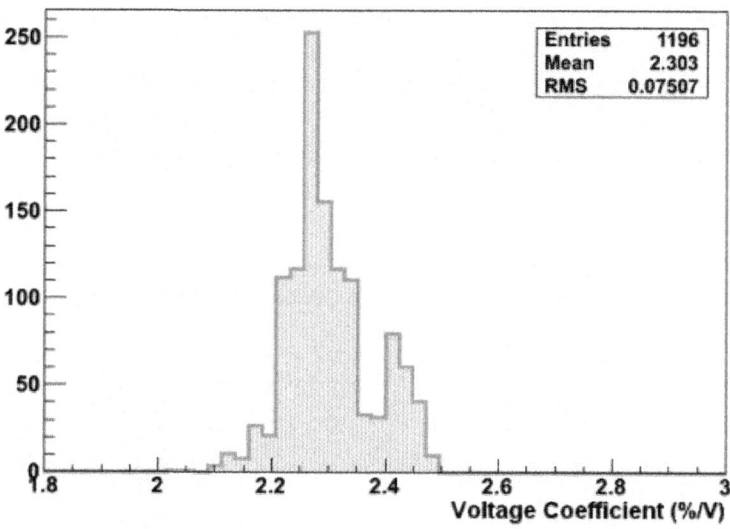

Figure 8: Statistical distribution of the voltage coefficients, for the same set of 1196 APDs.

EXTRACTION OF AMPLITUDE AND TIME INFORMATION: TRADITIONAL METHODS AND ALTERNATIVE APPROACHES

The output signal from Avalanche Photodiodes needs to be analyzed to extract as much as possible the information contained. Particularly relevant are of course the amplitude information, related to the amount of energy deposited in the individual module, and the timing information associated to it. The procedures to extract such information are not trivial, especially when analyzing events which span a large dynamical range, as it is the case for electromagnetic calorimeters in high energy experiments. In such a case, various algorithms have been developed and used, whose relative merits may be compared according to the precision and CPU time required. Even methods based on neural network topologies may be implemented and applied to simulated and real data. With reference to Fig.9, which shows a typical signal, as sampled by a flash ADC, the shape of the signal may be fitted by a Gamma function

$$ADC\ (t) = Pedestal + A^{-n}\ x^n\ e^{\ n(1-x)}\quad,\ x = (t-t_0)/\tau$$

(5)

where $\tau = n\ \tau_0$, τ_0 being the shaper constant, and n ~2. Such fit procedure is certainly able to provide reliable values of the amplitude A and time information t0 in case of large-amplitude signals, for which the number of time samples is relatively high (larger than 5-7). However, there are two main drawbacks inherent to this method: the algorithm is relatively slow, if one considers that it has to be applied to a large number of individual modules on an event-by-event basis, which is dramatic especially for on-line triggering. Secondly, in case of signals with very low amplitudes, the fit quite often provides unreliable values, since the signal shape is no longer similar to a Gamma function. For such reasons, alternative approaches have been tested and compared to the standard fitting procedure: fast fitting methods, peak analysis and so on. Here we want to show an example based on a neural network approach, which was recently tested on a sample of LED calibration data obtained for a large number (a few thousands) of channels in the ALICE electromagnetic calorimeter.

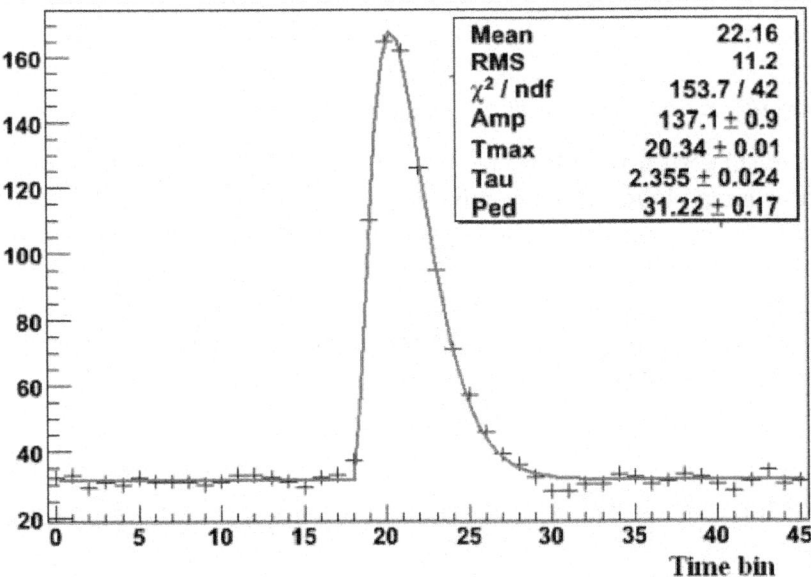

Figure 9: Shape of the signal, as extracted from a sampling ADC.

In order to prepare a data sample which exhibits its maximum at different times, as it could happen for real data, the LED signal was shifted in time every 100 events. Also the amplitude distribution is very broad, in order to span a region as large as possible, similarly to real data. This was due to the inevitable difference in the distribution of the light signal to the different modules. As a result, Figs.10 and 11 show two examples of a high amplitude (number of time samples = 12) and a low amplitude signal (number of time samples = 6). All the data were processed with the standard fit algorithm, which provided the reference for the learning phase in the neural network approach.

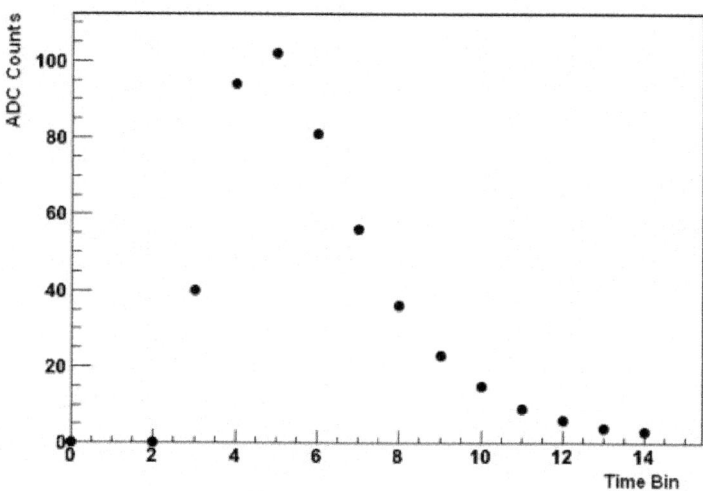

Figure 10: An example of a high amplitude signal, including 12 time samples

A feed forward multilayered neural network (Bishop, 1995) consists of a set of input neurons, one or more hidden layers of neurons, a set of output neurons, and synapses connecting each layer to the subsequent layer. The synapses connect each neuron in the first layer to each neuron in the hidden layer and each neuron of the hidden layer to the output (Fig.12). Several topologies may be chosen, as far as the number of input neurons and hidden layers are concerned.

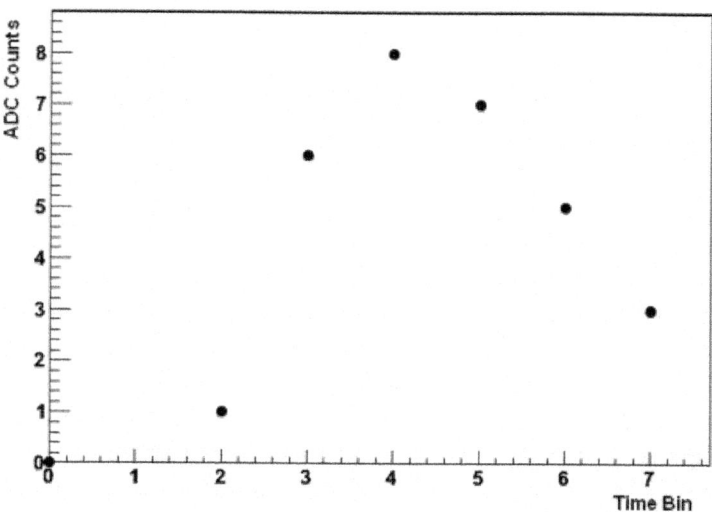

Figure 11: An example of a low amplitude signal, including only 6 time samples.

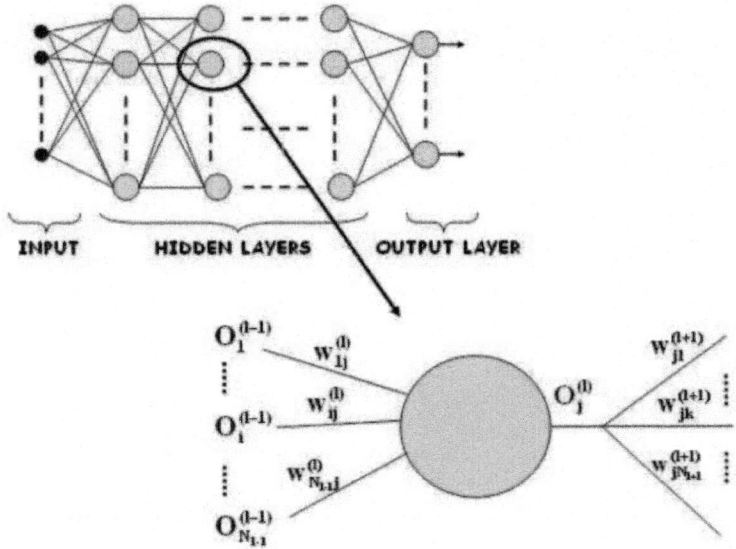

Figure 12: Schematic layout of a neural network.

The signal provided by the j-th neuron of the l-th layer is given by the linear combination of the neuron input values, where the w's are the weights:

$$a_j = \sum_{i=1}^{N_{l-1}} w_{ij}^{(l)} O_i^{(l-1)}$$

A backpropagation algorithm was used in the learning phase, in order to modify the initial values of the weights and minimize the error function:

$$E = \frac{1}{2} \sum_{p=1}^{N_P} \sum_{j=0}^{N_O} [y_{jp} - O_{jp}^N]^2$$

Best results were obtained in this case with 5 input neurons (the 5 values of the signal amplitude closest to the maximum), 10 hidden neurons and 2 output neurons (the amplitude and the time of the signal peak). Fig.13 shows the minimization of the error function with the number of epochs employed in the learning and testing phases. Figs.14 and 15 show the distributions of the differences between the reference values (provided by the Gamma-fit) and the output values from the neural network, both for the amplitude and the time. An RMS of 0.26 ADC channel was obtained for the signal amplitude, while a value of 0.007 channel bin (corresponding to 700 ps) was obtained for the time.

Such performance was compared to more traditional methods, based on fast fitting procedures or peak analysis methods, and it was shown that after a proper training phase, comparable results may be in principle obtained by a neural network, with a reduced CPU time.

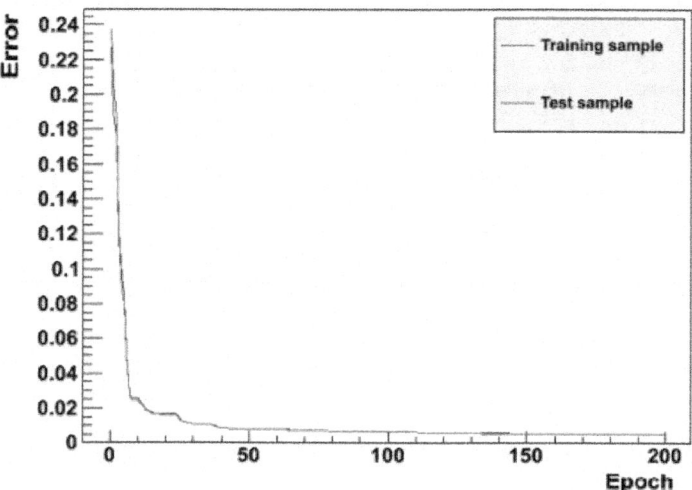

Figure 13: Minimization of the error function with a neural network.

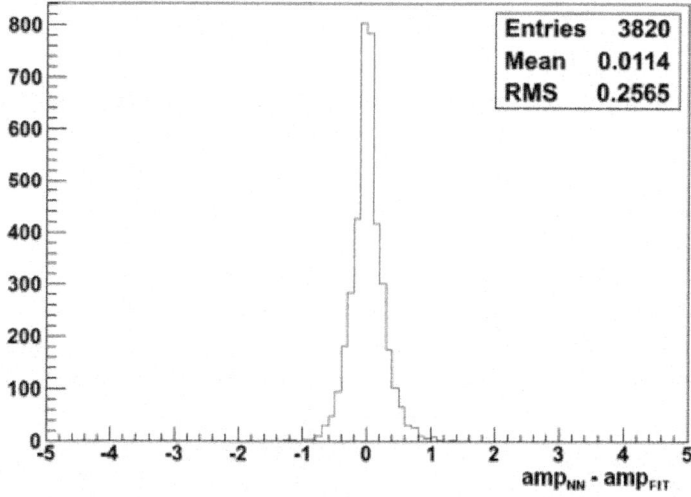

Figure 14: Distribution of the differences between the "true" value (provided by the fit with a Gamma-function) and the value provided by the neural network, in case of the signal amplitude.

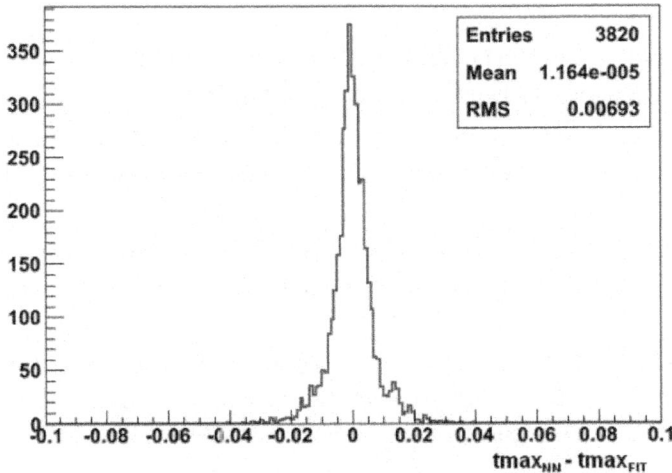

Figure 15: As for fig.14, for the time information.

CONCLUSION

After several years of R&D work, Avalanche Photodiodes have proved to be a mature technology to be routinely employed in the design and construction of large high-energy calorimeters for the readout of the scintillation light produced in the individual calorimeter cells. The use of APDs in high energy electromagnetic calorimetry has required large efforts from both physics Laboratories and Industries in order to improve several aspects allowing an efficient usage of these devices in particle detectors. As a result of these combined efforts, several devices have been developed which have a reasonable sensitive area, a suitable spectral sensitivity and a good resistance to radiations. Different experiments incorporating one or more electromagnetic calorimeters in their setup make now use of a large number (in the order of 105) of these devices with good results, and additional projects are looking forward to this solution. Several progresses are however possible along different directions. One aspect is certainly related to the increase in the sensitive area of the individual devices, without loosing any advantage originating from their intrinsic properties. This will allow a more efficient coupling of APDs to the scintillation crystals. Optimization of the spectral response in connection with the choice of the scintillation material is certainly another direction where some development could be expected in the next future. Additional improvements could come from the monitoring and control of such devices, in order to optimize and stabilize their gain as a function of the bias voltage and of the operating temperature.

REFERENCES

1.   Aamodt, K. et al., The ALICE Collaboration (2008). The ALICE detector at LHC, Journal of Instrumentation 3, S08002

2.   Anzivino, G. et al. (1995). Review of the hybrid photo diode tube (HPD) an advanced light detector for physics, Nuclear Instruments and Methods A365, 76-82

3.   Badalà, A. et al.(2008). Characterization of Avalanche Photodiodes for the electromagnetic calorimeter in the ALICE experiment, Nuclear Instruments and Methods A596, 122-125

4.   Badalà, A. et al.(2009). Prototype and mass production tests of avalanche photodiodes for the electromagnetic calorimeter in the ALICE experiment at LHC, Nuclear Instruments and Methods A610, 200-203

5.   Barlow, R.J. et al.(1999). Results from the BABAR electromagnetic calorimeter beam test, Nuclear Instruments and Methods A420, 162-180

6.   Bishop, C.M. (1995). Neural Networks for Pattern Recognition, Clarendon, Oxford Chartrchyan, S. et al., The CMS Collaboration (2008). The CMS detector at LHC, Journal of Instrumentation 3, S08004

7.   Erni, W. et al., The PANDA Collaboration (2008). Technical Design Report, arXiv :0810.1216v1

8.   Fabjan, C.W. & Gianotti, F. (2003), Review of Modern Physics 75,1243-1286

9.   Lorenz, E. et al. (1994), Fast readout of plastic and crystal scintillators by avalanche photodiodes, Nuclear Instruments and Methods A344, 64-72

10.   Karar, A.; Musienko, Yu. & Vanel, J.Ch. (1999), Characterization of Avalanche Photodiodes for calorimetry applications, Nuclear Instruments and Methods A428,413-431

11.   Musienko, Yu. (1992), The CMS electromagnetic calorimeter, Nuclear Instruments and Methods A 494, 308-312

12.   Wigmans, R. (2000), Calorimetry: Energy Measurements in Particle Physics, University Press, Oxford

Chapter 7

THE E-SCIENCE PARADIGM FOR PARTICLE PHYSICS

Kihyeon Cho

Korea Institute of Science and Technology Information Republic of Korea

INTRODUCTION

Research in the 21st century is increasingly driven by the analysis of large amounts of data within the e-Science paradigm. e-Science is the data centric analysis of science experiments unifying experiment, theory, and computing. According to Simon C. Lin and Eric Yen (Lin & Yen, 2009), e-Science or data-intensive science unifies theory, experiment, and simulations using exploration tools that link a network of scientists with their datasets. Results are analyzed using a shared computing infrastructure.

In this chapter, we use the concept of e-Science to combine experiment, theory and computing in particle physics in order to achieve a more efficient research process. Particle physics applications are generally regarded as a driver for developing this global e-Science infrastructure.

According to Tony Hey at Microsoft (Hey, 2006), thousands of years ago science focused on experiments to describe natural phenomena. In the last few hundreds of years, science became more theoretical. In the last few decades, science has become more computational, focusing on simulations. Today, science can be described as more data-intensive in nature, requiring a combination of experiment, theory, and computing. Attempts have been made to realize this e-Science concept. One e-Science application is the Worldwide Large Hadron Collider Computing Grid (WLCG), which realizes Ian Foster's definition of a grid (Foster et al., 2001). The grid is the combination of computing resources from multiple administrative domains to reach a common goal (Cho & Kim, 2009). As the global e-Science infrastructure is rapidly established, we must take advantage of worldwide e-Science progress. Highenergy physics has advanced the e-Science paradigm by successfully unifying experiments, theory, and computing (Cho et al., 2011).

We apply the e-Science concept to particle physics and show an example of this paradigm. As shown in Fig. 1, we construct a unified research model of experiment-theory-computing in order to probe the Standard Model and search for new physics. This is not a simple collection of experiments, computing, and theory, but a fusion of research in order to achieve a more efficient research process. We apply this concept to the Collider Detector at Fermi-lab (CDF) experiment in the USA and the Belle/Belle II experiment at High Energy Accelerator Research Organization (KEK) in Japan.

For computing-experiment, we construct and use the components of the e-Science research environment, including data production, data processing, and data analysis using collaborative tools. We also develop new computational tools for future experiments. In high energy physics, the goal of e-Science is to perform and/or analyze high energy physics experiments anytime and anywhere. We apply this system to the Belle II experiment at KEK. For data processing, WLCG is one of the original new research infrastructures that show how an effective collaboration might be conducted between users and facilities (Cho, 2007). The Asia Pacific area should develop both an e-Science platform and best practices for collaboration in order to fill the gaps in e-Science development between other continents. The Academia Sinica Grid Centre (ASGC), as the coordinator of the Asia federation under Enabling Grid in e-Science (EGEE), has worked closely with partners for region specific applications in data processing. For data analysis using collaborative tools, community building should be the foundation for collaboration rather than just offering technology. The e-Science research environment provides a trusted way to allow people, resources, and knowledge to connect and participate via a virtual organization. More and more countries will deploy a grid system and take part in the e-Science research environment. According to Simon C. Lin (Simon & Yen, 2009), we are widening the uptake of e-Science through close collaboration regionally and internationally.

For experiment-theory, we develop a combination of phenomenology and data analysis. Experiments give results and tools for theories and theories give feedback to experiments. We apply this system to the CDF, D0, and Belle experiments in order to probe the standard model and search for new physics. For theory-computing, we study lattice gauge theory and use the supercomputer at the Korea Institute of Science and Technology Information (KISTI).

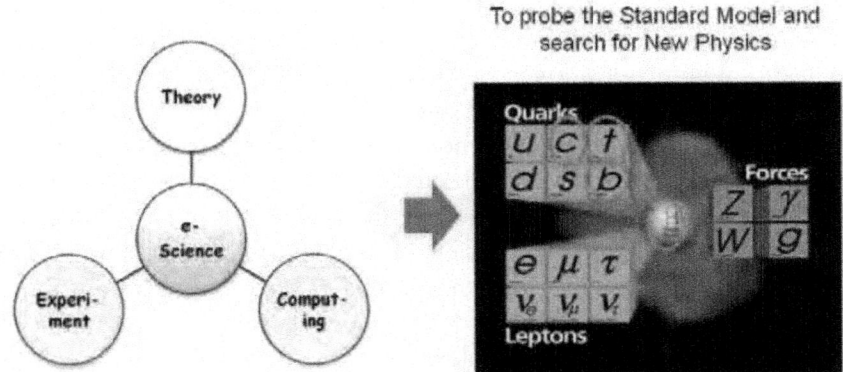

Figure 1: The paradigm of e-Science in high energy physics, which is a fusion of experiment, computing, and theory research.

MAIN

We explain the results for computing-experiment, experiment-theory, and theorycomputing for the analysis of particle physics. While many previous works have only used supercomputers, in our work computing results are combined with theory and experiment. We use a combination of supercomputers and an e-Science environment. The components of an e-Science environment are data production for remote shifts, data processing for grid farms, and data analysis using the Enabling Virtual Organization (EVO) collaborative tool.

For Computing-Experiment

E-Science research environment

We define a computing-experiment tool as an e-Science research environment. In order to study particle physics, we can access the environment anytime and anywhere even if we are not on-site an accelerator laboratory. A virtual laboratory enables us to perform research as if we were on-site (Cho, 2008). We apply e-Science components to the CDF experiment.

Data Production

The purpose of data production is to take both on-line shifts and off-line shifts anywhere. On-line shifts have been conducted through the use of a remote control room at KISTI and off-line shifts have been conducted via the sequential access through metadata (SAM) data handling (DH) system at KISTI. The remote control room is built to help non-US CDF members to fulfill their shift duties as a Consumer Operator (CO) part of the CDF data taking shift crew. The remote control room facilitates various monitoring applications that the CO has to monitor for a given eight hour shift. We have been operating the CDF remote control room at KISTI since July 22, 2008. A real Data Acquisition (DAQ) has been recorded at the remote control room at KISTI between August 1 and August 8, 2008. The CDF detector is an experimental apparatus for recording electrical events produced by the accelerator at an enormous rate. This apparatus is comprised of several components that perform different functions including a detector with millions of data channels transmitted to a corresponding number of electronic readout devices. The operation of an apparatus with this degree of complexity needs to be collaboratively controlled by researchers. In general, each shift crew takes an eight hour shift so that three shift crews will cover 24 hours. In the CDF experiment, the shift crew consists of three people with different missions. First, the Science Coordinator (SciCo) is responsible for the entire shift session and must have a lot of experience. The second person is the Ace shifter, who is an expert on the control of all detector components and electronic readout devices. The third person is the CO who has been trained in interpreting the meaning of the data being monitoring. UNIX processes intercept the on-line data transmitted from the front-end readout electronics and generate various plots that represent the quality of the data taken by the detector. These plots help the CO to determine whether or not the data collection is continuing as expected. Accordingly, the CO advises the Ace shifter to interrupt the detector operation in order to correct any problems.

Although the CO's monitoring task involves on-line data collection, this can be performed in a remote location due to its mostly monitoring-related nature. These remote control rooms are located at the Pisa University in Italy, the University of Tsukuba in Japan, and KISTI in Korea. In Korea, there are about 30 collaborators from six institutions, most of which have to fulfill CDF duties by taking detector operation shifts. All the plots that the consumers generate are accessible via web browsers where all the monitoring can be done. The CO has to not only monitor any plots generated by consumers but also must monitor the consumers themselves. However, the policy imposed by the Department of Energy (DOE) in the United States prohibits any remote researcher outside of Fermilab from executing any control-related UNIX command. Instead, control-related execution must be initiated by a person on-site. At the same time, all transmissions of control commands have to be encrypted using Kerberos. Thus, we can solve this problem by having an on-site crew send a graphic user interface (GUI) named "consumer controller" to the remote monitor via the Kerberized secure shell port. The CDF II experiment has been taking data from June 30, 2001 to September 30, 2011. Fig. 2 shows the CDF main operation center and remote control room at KISTI. As shown in Fig. 3, we have taken remote shifts (24 days per year on average) successfully.

Figure 2: The CDF main operation center and remote control room at KISTI.

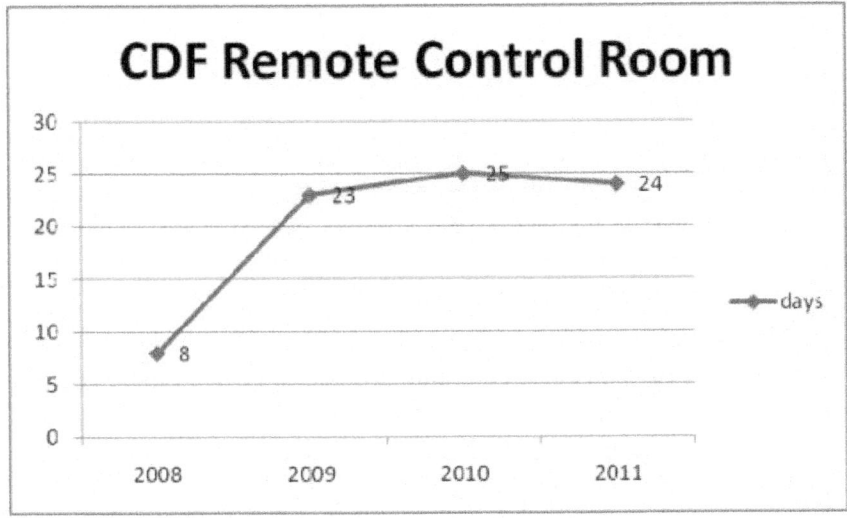

Figure 3: The CDF remote control used at KISTI.

We perform another type of remote data handling shift at KISTI. Whereas the remote control room implements an on-line version of remote data handling, there is a second shift that implements an off-line version of remote data handling. This second type of shift is actually in the form of a SAM DH shift. This shift also occurs eight hours per day for seven days. These shifts do not need to cover the entire twenty four hours with three shifts per day since they are off-line. Furthermore, one can take the shift in the daytime of his or her time zone if participating in the shift schedule outside of the USA. The CDF SAM DH is called off-line since the data handled in this case includes data inbound to the tape from SAM stations in reconstruction farms and vice versa. The off-line data transfers in CDF are between SAM stations and mass storage system (MSS). In Fermilab, MSS consists of a Storage Resource Manager (SRM), dCache, and the Enstore system. The dCache software was the result of joint project between Fermilab in Batavia, USA and DESY (Deutches Elecktronen SYnchrotron laboratory) in Hamburg, Germany. dCache is a front-end for disk caching and provides end-users with the functionalities of reading cached files and writing files to and from Enstore indirectly via dCache. The Enstore system is a direct interface to files on tape for end-users. End-users can refer to SAM stations of CAF and farm machines. In the present context, the SAM stations in the CDF Analysis Farm (CAF) and farm clusters use an Application Programming Interface (API) provided by dCache to read files from and write files to the tapes via dCache and the Enstore systems. Thus, the mission of the CDF SAM shift includes monitoring the Enstore system, the dCache system, and SAM stations of the CDF analysis farm (CAF) and the CDF experiment farm.

Data Processing

Data processing is accomplished using a High-Energy Physics (HEP) data grid. The objective of the high-energy physics data grid is to construct a system to manage and process highenergy physics data and to support the high-energy physics community (Cho, 2007).

For data processing, Taiwan has the only WLCG Tier-1 center and Regional Operation Center in Asia since 2005. ASGC has also been serving as the Asia Pacific Regional Operational Center to maximize grid service availability and to facilitate extension of eScience (Lin & Yen, 2009). In Japan, a Tier-2 computing center supporting the A Toroidal LHC Apparatus (ATLAS) experiment has been running at the University of Tokyo. There is another Tier-2 center at Hiroshima University for the A Large Ion Collider Experiment (ALICE) (Matsunaga, 2009). At KEK, collaborating institutes operate a grid site as members of the WLCG. These institutes try to use their grid resources for the Belle and Belle II experiments. The Belle II experiment, which will start in 2015, will use distributed computing resources.

We explain the history of data processing for the CDF experiment. The CDF is an experiment on the Tevatron, at Fermilab. The CDF group ran its Run II phase between 2001 and 2011. CDF computing needs include raw data reconstruction, data reduction, event simulation, and user analysis. Although very different in the amount of resources needed, they are all naturally parallel activities. The CDF computing model is based on the concept of a Central Analysis Farm. The increasing luminosity of the Tevatron collider has caused the computing requirement for data analysis and Monte Carlo production to grow larger than available dedicated CPU resources. In order to meet demand, CDF has examined the possibility of using shared computing resources. CDF is using several computing processing systems, such as CAF, Decentralized CDF Analysis Farm (DCAF), and grid systems. The Korea group has built a DCAF for the first time. Finally, we have constructed a CDF grid farm at KISTI using an LCG farm.

In 2001, we have built a CAF, which is a cluster farm inside Fermilab in the United States. The CAF was developed as a portal. A set of daemons accept requests from the users via kerberized socket connections and a legacy protocol. Those requests are then converted into commands to the underlying batch system that does the real work. The CAF is a large farm of computers running Linux with access to the CDF data handling system and databases to allow the CDF collaborators to run batch analysis jobs. In order to submit jobs we use a CAF portal with two special features. First, we can submit jobs from anywhere. Second, job output can be sent directly to a desktop or stored on a CAF File Transfer Protocol (FTP) server for later retrieval (Jeung et al., 2009).

In 2003, we have built a DCAF, a cluster farm outside Fermilab. Therefore, CDF users around the world enabled to use it like CAF at Fermilab. A user could submit a job to the cluster either at Central Analysis Farm or at the DCAF. In order to run the remote data stored at Fermilab in USA, we used SAM. We used the same GUI used in Central Analysis Farm (Jeung et al., 2009).

In 2006, we have built CDF grid farms in North America, Europe, and Pacific Asia areas. The activity patterns at HEP required a change in the HEP computing model from clusters to a grid in order to meet required hardware resources. Dedicated Linux clusters on the Farm Batch System Next Generation (FBSNG) batch system were used when CAF launched in 2002. However, the CAF portal has gone from interfacing to a FBSNG-managed pool to Condor as a grid-based implementation since users do not need to learn new interfaces (Jeung et al., 2009).

We have now adapted and converted out a workflow to the grid. The goal of movement to a grid for the CDF experiment is a worldwide trend for HEP experiments. We must take advantage of global innovations and resources since CDF has a lot of data to be analyzed. The CAF portal may change the underlying batch system without changing the user interface. CDF used several batch systems. The North America CDF Analysis Farm and the Pacific CDF Analysis Farm is a Condor over Globus model, whereas the European CDF Analysis Farm is a LCG (Large Hadron Collider Computing Grid) Workload Management System (WMS) model. Table 1 summarizes the comparison of grid farms for CDF (Jeung et al., 2009). Fig. 4 shows the CDF grid farm scheme (Jeung et al., 2009).

Users submit a job after they input the required information about the job into a kerberized client interface. The Condor over Globus model uses a virtual private Condor pool out of grid resources. A job containing Condor daemons is also known as a glide-in job. The advantage of this approach is that all grid infrastructures are hidden by the glide-ins. The LCG WMS model talks directly to the LCG WMS, also known as the Resource Broker.

This model allows us to use grid sites where the Condor over Globus model would not work at all and is adequate for grid job needs. Since the Condor based grid farm is more flexible, we applied this method to the Pacific CDF Analysis Farm (Jeung et al., 2009). The regional CDF Collaboration of Taiwan, Korea and Japanese groups have built the CDF Analysis Farm, which is based on grid farms. We called this federation of grid farms the Pacific CDF Analysis Farm.

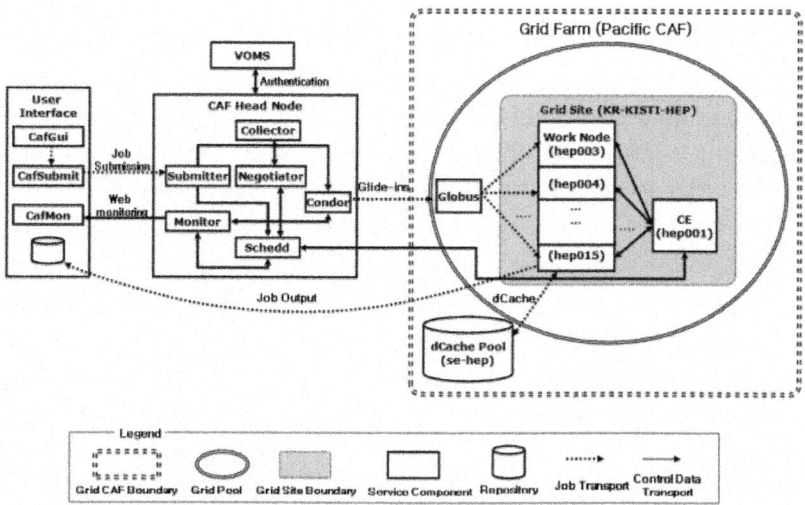

Figure 4: The scheme of the Pacific CDF analysis farm.

Table 1: Comparison of grid farms for CDF

Grid CDF Analysis Farm	Head node	Work node	Grid middle-ware	Method	VO (Virtual Organization)
North America CDF Analysis Farm	Fermilab (USA)	USSD (USA) etc	OSG	Condor over Globus	CDF VO
European CDF Analysis Farm	CNAF (Italia)	IN2P3 (France) etc	LCG	WMS (Workload Management System	CDF VO
Pacific CDF Analysis Farm	AS (Taiwan)	KISTI (Korea) etc	LCG, OSG	Condor over Globus	CDF VO

The Pacific CDF Analysis Farm is a distributed computing model on the grid. It is based on the Condor glide-in concept, where Condor daemons are submitted to the grid, effectively creating a virtual private batch pool. Thus, submitted jobs and results are integrated and are shared in grid sites. For work nodes, we use both LCG and Open Science Grid (OSG) farms. The head node of Pacific CDF Analysis Farm is located at the Academia Sinica in Taiwan. Now it has become a federation of one LCG farm at the KISTI in Korea, one LCG farm at the University of Tsukuba in Japan and one OSG and two LCG farms in Taiwan.

Data Analysis using Collaborative Tools

A data analysis using collaborative tools is for collaborations around the world to analyze and publish the results in collaborative environments. We installed an operator EVO server at KISTI. Using this environment, we study high energy physics for CDF and Belle experiments. EVO is the next version of its predecessor, Virtual Room Videoconferencing System (VRVS). The first release of EVO was announced in 2007. The EVO system is written in the Java programming language. The EVO system provides a client application named "Koala." The Koala plays two client roles in order to communicate with two types of servers. The first type is a central server located in Caltech and handles videoconferencing sessions. Participants can use a Koala to enter a session that another participant created or book a new session. Once a participant is in a session, the Koala starts to play the role of another type of client that now communicates with one of the networked servers that handle the flow of media streams. The second type of server comprising a network is called "Panda." When a Koala is connected to a specific Panda, the Koala initiates a video tool called "vievo" and an audio tool called "rat," both of which have their origins in the "MBone" project. EVO has improved upon VRVS with the following new features: support for Session Initiation Protocol (SIP), including ad-hoc or private meetings, encryption, private audio discussion inside a meeting, and whiteboard. In 2007, we constructed the EVO system at KISTI since the Korean HEP community is large enough to have its own EVO Panda servers. The configuration of two servers by the Caltech group enables the first Korean Panda servers to run. Fig. 5 shows communications between KISTI Panda servers and other Panda servers in the EVO network. Since its introduction in 2007, KISTI Panda servers have served many communities such as the Korean Belle community and the Korean CDF community.

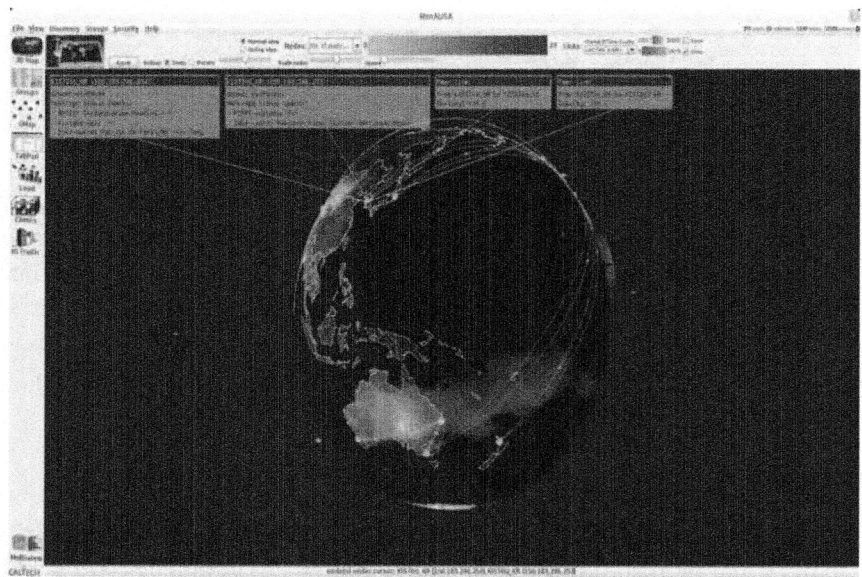

Figure 5: Communications between KISTI "Panda" servers and other "Panda" servers in the EVO network.

New computing-experimental tools

For new computing-experimental tools, we have worked on a Belle II data handling system. The Belle II experiment will begin at KEK in 2015. Belle II computing needs to include raw data reconstruction, data reduction, event simulation, and user analysis. The Belle II experiment will have a data sample about 50 times greater than that collected by the Belle experiment.

Therefore, we have very large disk space requirements and potentially unworkably long analysis times. Therefore, we suggested a meta-system at the event-level to meet both requirements. If we have good information at the meta-system level, we can reduce the CPU time required for analysis and save disk space.

The collider will cause the computing requirement for data analysis and Monte Carlo production to grow larger than available CPU resources. In order to meet these challenges, the Belle II experiment will use shared computing resources as the Large Hadron Collider (LHC) experiment has done. The Belle II experiment has adopted the distributed computing model with several computing processing systems such as grid farms (Kuhr, 2010).

In the Belle experiment (Abashian et al., 2002), we use a metadata scheme that employs a simple "index" file. This is a mechanism to locate events within a file based on predetermined analysis criteria. The index file is simply the location of interesting events within a larger data file. All these data files are stored on a large central server located at the KEK laboratory. However, for the Belle II experiment, this will not be sufficient as we will distribute the data to grid sites located around the world. Therefore, we need a new metadata service in order to construct the Belle II data handling system (Kim, et al. 2011; Ahn, et al., 2010).

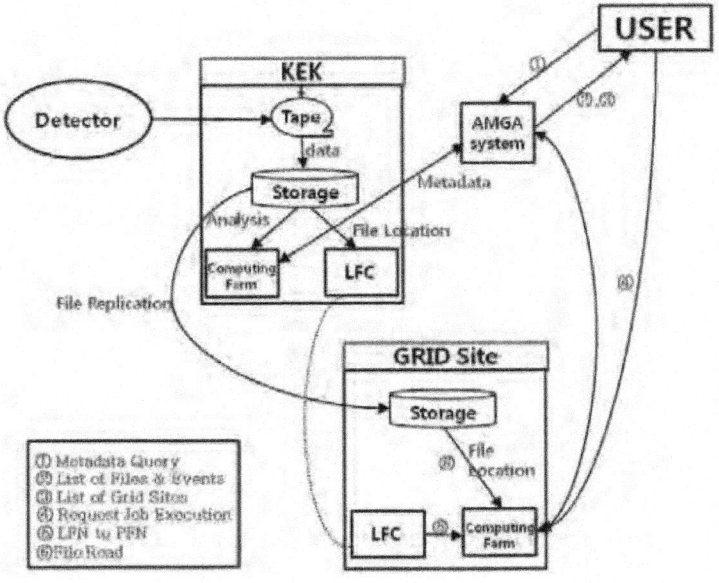

Figure 6: Data handling scenario at the Belle II experiment.

Fig. 6 shows the Belle II data handling system scheme. First, a user makes a metadata query to the server. Second, the server gives back a list of files and events. Third, the server may give a list of grid sites. Fourth, the user requests job execution at grid sites. Fifth, a logical file catalog (LFC) maps a logical file name (LFN) into a set of physical file names (PFN). Finally, the computing farms at the grid site read the requested physical file (Ahn, et al., 2011).

For Experiment-Theory

For experiment-theory, using the results of CDF and Belle experiments, we test phenomenological models of particle physics. Fig. 7 shows various physics topics for experiment-theory research, including Kaon Semi-leptonic form factor, rare B decay, mixing and CP (Charge Parity) violation on Bs→ J/ψ Φ, forward-backward asymmetry of top quarks, and CP violating dimuon charge asymmetry due to B mixing. Models for these physics topics include lattice gauge theory using staggered fermion, Left-Right models, and model-independent analysis. In this section, we introduce the left-right model and the forward-backward asymmetry of top quarks

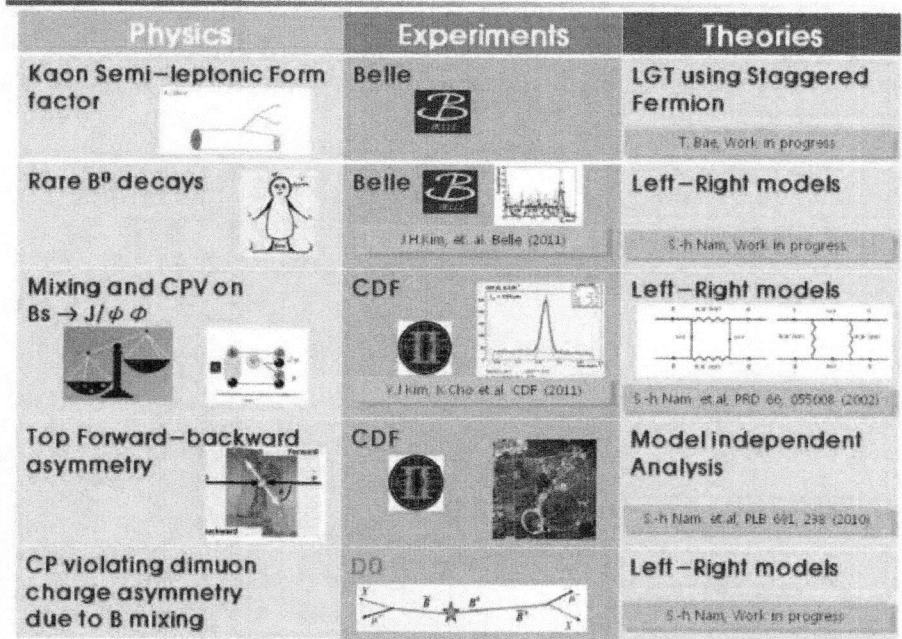

Figure 7: Physics topics related to experiment and theory.

Left- Right Model

In CDF experiments, we study mixing and CP violation on Bs→ J/ψ Φ decay channels. For this analysis, we apply Left-Right models and compare the results. We also apply to the same model to the CP violating dimuon charge asymmetry due to B mixing. Fig. 8 shows the feynman diagram of Left-Right models for the analysis of CP violating dimuon charge asymmetry due to B mixing.

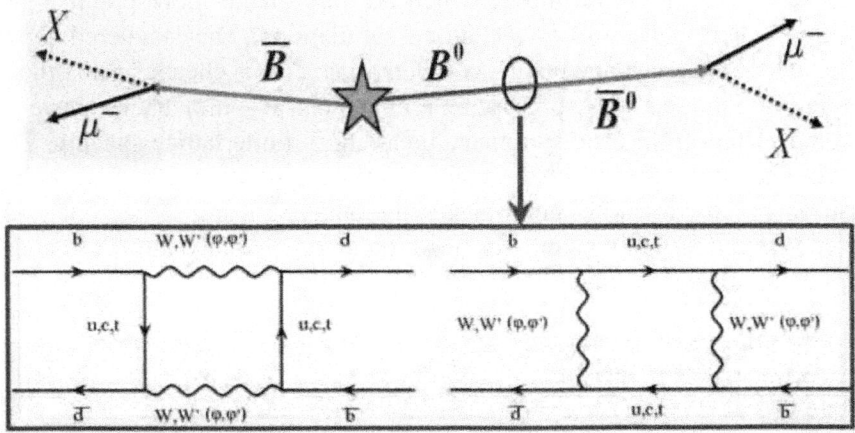

Figure 8: The Feynman diagram of Left-Right models for the analysis of CP violating dimuon charge asymmetry due to B mixing.

The forward-backward asymmetry of top quark pairs

In 2008, CDF showed a possible anomaly in the forward-backward asymmetry of the top quark, where AFB = 0.19± 0.07(stat.) ± 0.02(syst.) (Aaltonen et al., 2008). We have performed model independent analysis. Considering the s-, t-, and u- channel exchanges of spin-0 and spin-1 particles whose color quantum number is a singlet, octet, triplet or sextet, we study the region consistent with the CDF data at a one sigma level. We show the necessary conditions for the underlying new physics in a compact and effective way when those new particles are too heavy to be produced at the Tevatron. However, the results still affect the forward-backward asymmetry of top quark.

For theory-computing

For theory-computing, we study flavor physics based on lattice gauge theory, which enables large-scale numerical simulations on a supercomputer. The theory of strong interactions in the Standard Model is Quantum Chromo Dynamics (QCD). In phenomena related to the Cabibbo-Kobayashi-Maskawa (CKM) matrix, the theoretical values of the interaction amplitudes also have factors that cannot be obtained in a perturbative way since the strong coupling constant becomes strong at a low energy scale as QCD, as a non-abelian gauge theory, predicts. The only way that one can calculate the non-perturbative quantities with a controlled error is the lattice method, in which we put strongly interacting particles, quarks and gluons, on a lattice and calculate quantities directly from first principles. Fig. 9 shows the baryon based on lattice QCD.

We use the staggered fermions, which are one of the more popular lattice fermion schemes for full QCD lattice simulations. The staggered fermion scheme has the advantage that its computational cost is cheaper than other lattice fermion models while preserving remnant chiral symmetry. However, this scheme suffers from taste symmetry breaking in finite lattice spacing. Tastes are the remaining species that originate from the fermion doubling problem. Taste symmetry breaking complicates the analysis using lattice data. Thus, in order to reduce taste symmetry breaking effects, we use the HYP-smeared staggered fermions as valence quarks.

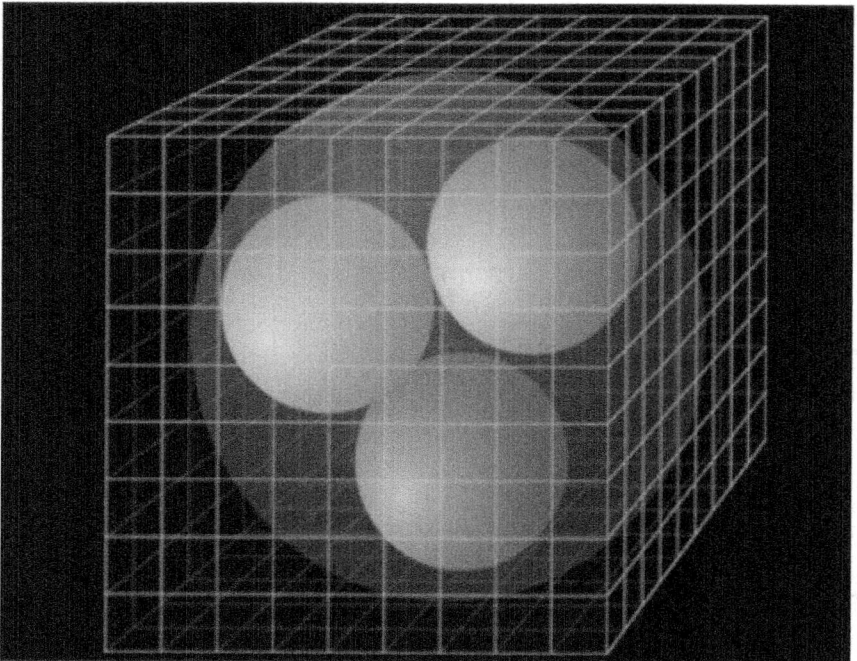

Figure 9: Baryon based on lattice QCD.

Lattice calculations cannot be done in the physical quark mass regime. In order to overcome this limitation, we calculate quantities with several non-physical quark masses and extrapolate the result to a physical regime. In this procedure, the staggered chiral perturbation theory guides the extrapolation. This study can be extended to heavy flavor physics and other hadronic phenomena. In addition to physics research, we have developed new algorithms that enhance precision and utilize new hardware such as Graphic Processing Unit (GPU), which overcomes the limitation of CPU computing power.

Kaon semi-leptonic decay form factor

Fig. 10 shows the diagram for kaon semi-leptonic decay. The CKM matrix elements are quark mixing parameters, which can be determined by combining experimental weak decay widths of hadrons and their theoretical calculations. A traditional way to determine Vus is connected with the kaon semi-leptonic decay channels, which include $K^+ \to \pi^0\, l^+\, \nu_l\, (K^+_{l3})$ and $K^0 \to \pi^-\, l^+\, \nu_l\, (K^0_{l3})$. Using these types of decays, we use the conserved vector current operator and the scalar density operator.

The decay rate of K_{l3} is written as the product of $|V_{us}|^2$ and $|f_+(0)|^2$. The vector form factor at zero momentum transfer, $f_+(0)$, is defined from the hadronic matrix element of the vector current between kaon and pion states. The matrix elements of the vector current can be extracted from the three-point correlation function whose interpolating operators are composed by the pseudo-scalar operator and the conserved vector current operator.

In this method, we have to generate quark propagators first. In order to create the desired meson states (kaon or pion) with non-zero spatial momenta, we use random U(1) sources with momentum phases. We also use the PxP operator insertion method (generally called sequential source) in order to create or annihilate the other meson state. Next, we contract these quark propagators properly and obtain three-point correlation function data.

From a Ward identity, we can convert the matrix elements of the vector current operator to those of the scalar density operator. This gives another method to calculate the form factor. The way to obtain correlation function data is similar to that found for the vector current method. Since the two methods are connected by a Ward identity, we can check if the data is consistent

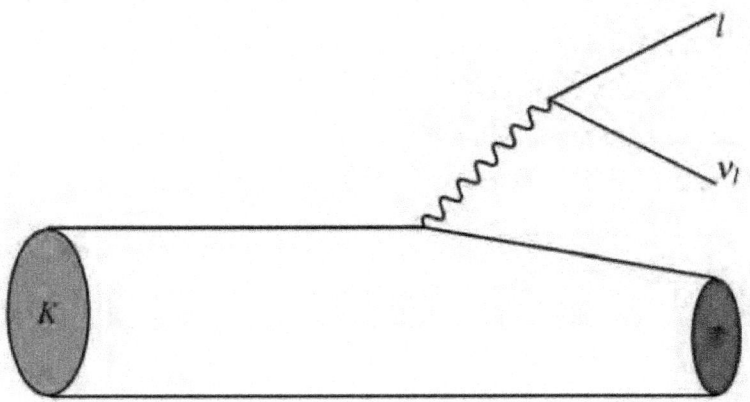

Figure 10: Kaon semi-leptonic decay.

Kaon and pion decay constants

The kaon and pion decay constants can also be used to determine V_{us}. Since the ratio $f_K/f\pi$ is related to V_{us}/V_{ub}, we can obtain V_{us} if V_{ub} is precisely known. From these quantities, we calculate the two point function of axial vector current and pseudo-scalar operator in the same way as the form factor.

CONCLUSIONS

We have introduced the concept of an e-Science paradigm for experiment-computing-theory for particle physics. Computing-experiment collaborative research offers not only an e-Science research environment including data production, data processing and data analysis, but also a data handling system for the Belle II experiment. The e-Science research environment enables us to research particle physics anytime and anywhere in more efficient way. Experiment-theory collaborative research provides a way to study the standard model and new physics. Theory-Computing collaborative research enables lattice gauge theory tools using supercomputing at KISTI.

In conclusion, we presented a new realization of e-Science paradigm of experiment, theory and computing in particle physics. Applying this concept to particle physics, we can achieve more efficient results to test the standard model and search for new physics.

ACKNOWLEDGMENT

I would like to express my thanks to the members of high energy physics team (Junghyun Kim, Soo-hyeon Nam, Youngjin Kim and Taegil Bae) at KISTI for the work.

A GLOSSARY OF ACRONYMS

ALICE: A Large Ion Collider Experiment

API: Application Programming Interface

ASGC: Academia Sinica Grid Centre

ATLAS: A Toroidal LHC Apparatus

CAF: CDF Analysis Farm

CDF: Collider Detector at Fermilab

CKM: Cabibbo-Kobayashi-Maskawa

CO: Consumer Operator

CP: Charge-Parity

DAQ: Date Acquisition

DCAF: Decentralized CDF Analysis Farm

DESY: Deutches Elecktronen SYnchrotron laboratory

DH: Data Handling

DOE: Department of Energy

GUI: Graphic User Interface

EGEE: Enabling Grid in e-Science

EVO: Enabling Virtual Organization

FBSNG: Farm Batch System Next Generation

FTP: File Transfer Protocol

GPU: Graphic Processing Unit

HEP: High-Energy Physics

KEK: High Energy Accelerator Research Organization in Japan

KISTI: Korea Institute of Science and Technology Information

LCG: Large Hadron Collider Computing Grid

LFC: Logical File Catalog

LFN: Logical File Name

LHC: Large Hadron Collider

MSS: Mass Storage System

OSG: Open Science Grid

PFN: Physical File Name

QCD: Quantum Chromo Dynamics

SAM: Sequential Access through Metadata

SciCo: Science Coordinator

SIP: Session Initiation Protocol

SRM: Storage Resource Manager

VRVS: Virtual Room Videoconferencing System

WLCG: Worldwide Large Hadron Collider Computing Grid

WMS: Workload Management System

REFERENCES

1.   Aaltonen, T. et al. [CDF Collaboration] (2008). Forward-Backward Asymmetry in Top-Quark Production in ppbar Collisions at √s=1.96TeV, Physical Review Letters, Vol. 101, No. 20, pp. 202001, 0031-9007.

2.   Abashian, A. et al [The Belle Collaboration] (2002). The Belle Detector, Nuclear Instruments and Methods in Physics Research Section A: Accelerators, Spectrometers, Detectors and Associated Equipment, Vol. 479, No. 1, pp. 117-232, 0168-9002.

3.   Ahn. S., et al. (2010). Design of the Advanced Metadata Service System with AMGA for the Belle II Experiment, Journal of Korean Physical Society, Vol. 57, No. 4, pp. 715-724, 1976-8524.

4.   Ahn, S., Kim, J. H., Huh, T., Hwang, S., Cho, K. et al (2011). The Embedment of a Metadata System at Grid Farms at the Belle II Experiment. Journal of Korean Physical Society, Vol. 59, No. 4, pp. 2695-2701, 1976-8524.

5.   Cho, K. (2007). Cyberinfrastructure in Korea, Computer Physics Communications, Vol. 177, No. 1-2, pp 247-248, 0010-4655.

6.   Cho K. (2008). e-Science for High Energy Physics in Korea, Journal of Korean Physical Society, Vol. 53, No. 92, pp.1187-1191, 1976-8524.

7.   Cho, K. & Kim, H. W. (2009) Heavy Flavor Physics through e-Science, Journal of Korean Physical Society, Vol. 55, No. 52, pp. 2045-2050, 1976-8524.

8.   Cho, K., Kim, H., & Jeung, M. (2010). Cyberinfrastructure for High Energy Physics in Korea, Journal of Physics: Conference Series, Vol. 219, No. 7, 072032, 1742-6596.

9.   Cho, K., Kim, J. H., & Nam, S-H., (2011). Collider physics based on e-Science paradigm of experiment-computing-theory. Computer Physics Communications, Vol. 182, No. 9, pp. 1756-1759, 0010-4655.

10.   Foster, I., Kesselman C., & Tuecke, S. (2001). The Anatomy of the Grid: Enabling Scalable Virtual Organizations, International Journal of High-Performance Computing Applications, Vol. 15, No. 3, pp. 200-222, 1094-3420.

11.   Hey, T. (2006), e-Science and Cyberinfrastructure. Keynote lecture at the 20th International CODATA Conference, Beijing, China. 24 October 2006.

12.   Jeung, M. et al. (2009). The Data Processing of e-Science for High Energy Physics, Journal of Korean Physical Society, Vol. 55, No. 52, pp 2067-2071, 1976-8524.

13. Kim, J. H. et al. (2011). The advanced data searching system with AMGA at the Belle II experiment, Computer Physics Communications, Vol. 182, No. 1, pp. 270-273, 0010-4655.

14. Kuhr, T. (2010). Computing at Belle II, Proceedings of Computing in High Energy Physics 2010, Taipei, October 2010.

15. Lin, S. C. & Yen, E. (2009). e-Science for High Energy Physics in Taiwan and Asia. Journal of Korean hysical Society, Vol. 55, No. 52, pp.2035-2039, 1976-8524.

16. Matsunaga, H. (2009). Grid Computing for High Energy Physics in Japan. Journal of Korean Physical Society, Vol. 55, No. 52, pp.2040-2044, 1976-8524.

Chapter 8

MUON COLLIDERS AND NEUTRINO EFFECTIVE DOSES

Joseph John Bevelacqua

Bevelacqua Resources USA

INTRODUCTION

Lepton accelerators incorporate electron, muon, and tau beams. First generation lepton machines, electron accelerators, are basic research tools and their radiation characteristics are well established. A second generation muon machine presents additional research possibilities as well as new health physics challenges. Third generation tau accelerators are currently theoretical abstractions and little development has been forthcoming. Although this chapter focuses on muon colliders and their unique radiation characteristics, initial scoping calculations for tau colliders are presented.

Neutrinos are electrically neutral particles, interact solely through the weak interaction, and have very small interaction cross sections (Particle Data Group 2010). They are present in the natural radiation environment due to cosmic rays, solar and terrestrial sources, and are produced during fission reactor and accelerator operations. From a health physics perspective these neutrino sources produce effective doses that are inconsequential. Although this will remain true for a number of years, planned muon accelerators or colliders will produce copious quantities of TeV energy neutrinos. In the TeV energy region, the health physics consequences of neutrinos can no longer be ignored. Upon operation of these accelerators, neutrino detection and the determination of neutrino effective doses will no longer be academic exercises, but will become practical health physics issues.

In a muon collider, neutrinos are produced when muons decay. The neutrino effective dose arises from neutrino interactions that produce showers or cascades of particles (e.g., neutrons, protons, pions, and muons). It is the particle showers that produce the dominant contribution to the neutrino effective dose (Bevelacqua, 2004).

Concerns for consequential neutrino effective doses have been previously postulated. Collar (1996) presented a hypothesis that the final stages of stellar collapse could produce neutrino effective doses that are sufficiently large to lead to the extinction of some species on earth. This concern has been challenged (Cossairt et al., 1997; Cossairt & Marshall, 1997), but the potential concern for large neutrino effective doses, on the order of hundreds of mSv/y or greater, remains, particularly for the planned muon colliders that will become operational in the next few decades of the 21st Century (Autin et al., 1999; Bevelacqua, 2004; Geer, 2010; King, 1999a; Kuno, 2009; and Zisman, 2011).

As background for muon colliders, an overview of the radiation environment at an electron accelerator is presented. This overview provides a foundation for a discussion of the characteristics of muon decays and the resultant neutrino effective doses. The characteristics of muon accelerators are addressed in this chapter and models for calculating the neutrino effective dose at a muon collider are provided. The radiological impacts of muon colliders and how basic dose reduction principles are affected by the underlying physics inherent in weak interaction processes are also discussed. Finally, a brief discussion of the neutrino effective doses anticipated at a third generation tau collider is provided.

ELECTRON-POSITRON COLLIDERS

Although this chapter addresses the neutrino effective dose from a muon collider, it is illustrative to provide a summary of the effects of other radiation types within a lepton collider (Bevelacqua, 2008, 2009, 2010a). These radiation fields are illustrated by considering an electron-positron collider. The radiation field within the muon collider facility is similar to those described in this section for electron-positron colliders.

An electron-positron collider accelerates electrons and positrons in circular rings before colliding the individual beams. There are a number of electron-positron colliders that have operated, are currently operating, or are being planned. These include the Large Electron Positron (LEP) Collider, and other machines summarized in the Review of Particle Properties (Particle Data Group, 2010). A new electron-positron machine, the International Linear Collider, is under design and is addressed from a health physics perspective in Bevelacqua (2008).

From an experimental physics perspective, electron-positron colliders have a number of advantages when compared to hadron colliders. First the collision results are less complex in terms of the particles produced, because electrons and positrons are fundamental particles without underlying structure or features. Hadrons are composed of quarks, but the electron and positrons have no such substructures. Therefore, the lepton's final state interactions are less complex than the structures that are produced from the interaction of the hadron's quarks. Particle interaction complexity is not the only advantage of electronpositron colliders.

The lepton colliders are also capable of achieving larger luminosities than hadron colliders. In addition, an order of magnitude less energy is required in electron-positron machines vice hadron colliders to achieve similar experimental results. For example, an electronpositron collider with a center-of-mass energy of 2 TeV is roughly equivalent to a 20 TeV center-of-mass energy hadron collider. In spite of these advantages, electron-positron collider health physics concerns exist (Bevelacqua; 2008, 2009, 2010a).

Electron-positron colliders produce more bremsstrahlung than hadron colliders. This bremsstrahlung production serves to limit the upper energies achieved by circular electronpositron colliders. In addition, electric power requirements rapidly increase with increasing energy unless beam power recovery mechanisms are developed and implemented. The bremsstrahlung produced in a circular electron-positron collider is a fundamental concern that can only be decreased by increasing the circumference of the machine. The logical conclusion is to use an accelerator with an infinite radius (i.e., a linear collider). This is most easily achieved by replacing the dual beams in a circular collider with colliding beams from two linear colliders.

The electron and positron beams produce a variety of radiation types that are derived from the direct beam and its interactions. Secondary radiation is produced from bremsstrahlung when beam particles strike accelerator components and from synchrotron radiation when beam particles are defected by magnetic fields.

Bremsstrahlung has a number of health physics consequences. These health physics issues include (NCRP 144, 2003): (1) electromagnetic cascade radiation containing high-energy photons, electrons, and positrons, (2) high-energy radiation including neutrons, pions, muons, and other hadrons, (3) activation of accelerator structures and components, (4) activation of air, cooling water, and soil, and (5) ozone and oxides of nitrogen produced in the air. Synchrotron radiation also has health physics consequences including: (1) electromagnetic cascade radiation, (2) photons, (3) neutrons, (4) activation of accelerator structures and components, (5) activation of air, cooling water, and soil, and (6) ozone and oxides of nitrogen produced in the air. These secondary radiation categories and their health physics consequences are addressed in more detail in subsequent discussion and in Bevelacqua (2008, 2009, 2010a).

The primary electron (positron) beams are contained within beam tubes, and secondary radiation is produced when the primary particles exit the beam tube either by design or accident. When electrons (positrons) exit the beam tube they strike accelerator components such as the beam tube structure, vacuum components, collimators, or structural members. When this occurs, the beam particle decelerates and radiates photons through the process of bremsstrahlung. The high-energy, bremsstrahlung photons produce electron-positron pairs that lead to additional bremsstrahlung. This process repeats itself, and produces an electromagnetic shower or cascade that contains numerous particles and a spectrum of photons having energies up to the kinetic energy of the initial beam particles

A second category of secondary radiation occurs when the beam particles traverse the accelerator's magnetic fields. The magnetic field produces a force that alters the particle's trajectory. It also changes the particle's velocity and leads to the emission of photon radiation. This process is known as synchrotron radiation. Synchrotron radiation is related to bremsstrahlung because a change in velocity or acceleration is involved in both processes. However, the synchrotron radiation differs from the bremsstrahlung spectrum.

With bremsstrahlung, the photon energy extends from zero up to the energy of the beam particle. However, synchrotron radiation is governed by the configuration and strength of the magnetic field. Therefore, the synchrotron spectrum is machine specific. For example, CERN's decommissioned Large Electron-Positron collider had a synchrotron spectrum that extended from the range of visible light to a maximum intensity that occurred in the range of a few hundred keV (Bevelacqua, 2008). The synchrotron radiation intensity rapidly decreases from its peak value as the photon energy increases above a few MeV. Both bremsstrahlung and synchrotron radiation induce an electromagnetic cascade.

The net result of the electromagnetic cascade is the deposition of energy in materials that are penetrated. This energy includes both particles stopped in the material and photon absorption. The photons produce additional secondary radiation and particles (e.g., photoneutrons) that activate accelerator materials. These same mechanisms lead to effective doses when personnel are in the presence of this radiation. These secondary radiation types are usually attenuated to insignificant levels by the concrete and earth shielding outside the accelerator tunnels containing the beam tubes. From a health physics perspective, the energy loss of the circulating, accelerating electrons and positrons produces synchrotron radiation (photons). Given the mass of the electrons and positrons, their trajectories are easily altered. Therefore, synchrotron radiation is expected to be a large fraction of the available beam power.

The synchrotron radiation requires shielding, and the extent of the shielding depends on the specific location within the accelerator facility. The amount of synchrotron radiation depends on the specific design characteristics of the electron-positron collider. Dominant factors governing the production of synchrotron radiation are the beam power and radius of curvature of the accelerator ring. From a practical standpoint, radiation generated from the circulating electron and positron beams occurs within the unoccupied shielded ring and is not normally a health physics issue.

The dominant contributors to the radiation environment at an electron-positron facility include electromagnetic cascade showers, external bremsstrahlung, photoneutrons, muons, and synchrotron radiation. Muon pair production in the Coulomb field of a nucleus is possible above a photon energy of about 211 MeV. This process is analogous to electronpositron pair production, but the muon pair production cross-sections are smaller by a factor of about 40,000 due to the differences in electron (0.511 MeV) and muon (105.7 MeV) masses (Bevelacqua, 2008).

The dominant muon pair production process is coherent muon production. In coherent production, the target nucleus remains intact as it recoils from the photon interaction. In a few percent of the time, the nucleus breaks-up with the resultant emission of muons. Muons also result from the decay of photopions and photokaons. However, the number of muon decays in a conventional electron-positron collider is not sufficient to produce a neutrino effective dose concern. To understand the neutrino effective dose from a muon collider, it is necessary to understand neutrino physics and neutrino interactions.

BASIC NEUTRINO PHYSICS

The current view of elementary particle physics is embodied in the Standard Model of Particle Physics (Cottingham & Greenwood, 2007; and Griffiths, 2008) that assumes all matter is composed of three types of fundamental or elementary particles: leptons, quarks, and mediators of the fundamental interactions. Bevelacqua (2010b) provides a description of the Standard Model from a health physics perspective. Leptons interact primarily through the weak interaction and electrically charged leptons also experience the effects of the electromagnetic force. They are not affected by the strong interaction. The leptons may be naturally grouped into three families or generations as (e^-, v_e), (μ^-, v_μ), and (τ^-, v_τ).

Neutrinos are neutral leptons, once believed to be massless, but now evidence suggests they have a non-zero mass (Particle Data Group, 2010).

The electron and muon neutrinos are well studied, but less is known about tau neutrinos.

To allow for massive neutrinos, the Standard Model must be modified and its assumptions altered. However, current experimental knowledge of neutrino properties does not permit the selection of a specific modification to the model. For example, it is not known if neutrino masses are to be interpreted as evidence of new, light, fermionic degrees of freedom (e.g., Dirac neutrinos), new, heavy, degrees of freedom (e.g., Majorana neutrinos), or whether a more complicated electroweak-symmetry-breaking interaction is present. However, the Standard Model is sufficient for the purposes of this chapter.

Within the Standard Model, neutrino effective doses are determined from the muon decay processes:

$$\mu^- \rightarrow e^- + \nu_\mu + \bar{\nu}_e \tag{1}$$

$$\mu^+ \rightarrow e^+ + \bar{\nu}_\mu + \nu_e \tag{2}$$

The neutrino effective doses depend on the number of muon decays, and the subsequent production of neutrinos. Specific effective dose relationships are provided in subsequent discussion.

NEUTRINO INTERACTIONS RELATED TO EFFECTIVE DOSE

In a muon collider, muon decays arise principally from Eqs. 1 and 2 that produce neutrinos and antineutrinos. The neutrinos interact through a variety of complex processes. A neutrino interaction discussion is simplified by following the methodology of Cossairt et al. (1997) and defining four processes (A, B, C, and D) to describe neutrino interactions with matter. The deposition of energy into tissue defines the effective dose (Bevelacqua, 2009, 2010a).

Process A involves neutrino scattering from atomic electrons. Electrons that recoil from elastic neutrino scattering deposit their energy in tissue and produce a neutrino effective dose. Process A occurs over a wide range of energy and the electron tissue interaction may involve multiple scattering of electrons.

In Process B, neutrinos interact coherently with nuclei. This process is only effective for low neutrino energies where the neutrino wavelength is too long to resolve the individual nucleons within the nucleus. At higher energies, Processes C and D become more important. Process B leads to low-energy ions having large linear energy transfer values. These ions deposit their energy into tissue according to their ranges, which are typically $\ll 1$ cm. Although Process

B is independent of the neutrino generation, the cross section for neutrinos is about twice the antineutrino cross section (King 1999a

Process C involves neutrino scattering from nucleons without shielding between the neutrinos and tissue. At energies below about 500 MeV, tissue dose is due to recoil nucleons. As the neutrino energy increases above about 0.5 GeV, secondary particle production increases. Eventually, these secondary particles produce particle showers or cascades in tissue. Process C is independent of the neutrino generation, affecting all three generations in the same manner.

Process D is similar to Process C with the exception that the neutrinos are shielded before striking tissue. Neutrinos with energy greater than about 0.5 GeV, emerging from a layer of material (e.g., earth shielding), result in a larger effective dose than unshielded neutrinos. The increase in effective dose arises from the fact that the tissue is exposed to the secondary particles produced by neutrino interactions in the shielding material as well as the neutrino beam. Process D is also independent of the neutrino generation.

A process that involves an increase in effective dose with added shielding is unique. One of the basic tenants for reducing effective dose for most radiation types (e.g., alpha and beta particles, heavy ions, muons, neutrons, photons, peons, and protons) is shielding the radiation source (Bevelacqua, 2009 and 2010a). The unique nature of Process D has a significant impact on the evaluation and control of neutrino effective dose.

NEUTRINO BEAM CHARACTERISTICS AT A MUON COLLIDER

Neutrinos are produced when the muon beam particles decay (See Eqs. 1 and 2). Weak interactions of muon neutrinos can be described in terms of two broad categories: charged current and weak current interactions. Charged current interactions involve the exchange of W-bosons to form secondary muons. Neutral current interactions produce uncharged particles through the exchange of Z-bosons. Both types of interactions produce hadron particle showers. Therefore, the neutrino induced radiation hazard will include secondary muons and hadronic showers. The hadronic showers have a much shorter range than the muons, but the number of particles in a hadronic shower can be quite large. The neutrino radiation hazard arises from these penetrating charged particle showers (Bevelacqua, 2008).

For TeV energy neutrinos, direct neutrino interactions in man account for less than 1% of the total effective dose because the primary hadrons from the neutrino interactions will typically exit the person before producing a charged particle shower (King, 1999b; Cossairt et al., 1996, 1997). Most of the neutrino effective dose is derived from particle showers produced in the shielding material.

The muon beam and subsequent neutrino beam are assumed to be well-collimated and to have a minimum divergence angle. For practical situations, the muons in the accelerator beam will have a small divergence angle and will be periodically focused using electromagnetic fields to ensure their collimation. No beam divergence is assumed in the subsequent calculations. Therefore, the actual beam will be somewhat more diffuse than assumed in the neutrino effective dose calculations. The neutrino beam will still produce particle showers, but they will be somewhat broader and less intense than the assumed well-collimated result. The beam divergence is analogous to the divergence of a laser beam as it exits an aperture (Bevelacqua, 2009, 2020).

The magnitude of the effective dose from a particle shower is dependent on the material in the interaction region lying directly upstream of the individual being irradiated. Calculation of the neutrino effective dose considers the configuration where a person is (1) completely bathed in the neutrino beam, and (2) is surrounded by material that will produce particle showers from neutrino interactions. These requirements lead to a bounding set of effective dose predictions.

These assumptions are too conservative for the TeV energies that will be encountered in mature muon colliders, but they provide a bounding neutrino effective dose result given the current level of design. Basic physics principles suggest that the neutrino interactions will be more peaked in the beam direction as the muon energies increase. In addition, the neutrino beam radius (r) will be relatively small and is given by (King, 1999b):

$$r = \theta L \tag{3}$$

where θ is called the characteristic angle, opening half-angle, or half-divergence angle of the muon decay cone

$$\theta = \frac{m\,c^2}{E} \tag{4}$$

In Eqs. 3 and 4, L is the distance to the point of interest such as the distance from the muon decay location to the earth's surface, θ is given in radians, E is the muon beam energy, and mc^2 is the rest mass of the muon (105.7 MeV). As the muon energy increases, the neutrino beam radius and size of the resultant hadronic showers tend to be smaller than the size of a person.

The characteristic angle varies inversely with energy. If E is expressed in TeV:

$$\theta \approx \frac{10^{-4}}{E[TeV]}$$

$$(5)$$

Therefore, the emergent neutrino beam will consist of a narrow diverging beam that is conical in shape.

Table 1 summarizes straw-man muon collider parameters (King, 1999b). It should be noted that the straw-man muon colliders are constructed below the earth's surface to provide muon shielding. However, the neutrino attenuation length is too long for the beam to be appreciably attenuated by any practical amount of shielding, including the expanse of ground between the collider and its exit from the surface of the earth. Therefore, the effective dose reduction principle as applied to neutrinos will no longer include shielding as an element. In fact, shielding the neutrino beam will produce hadronic showers and increase the effective dose. This peculiar behavior has its basis in the nature of the weak interaction, the uncharged nature of the neutrino, and the TeV energies that will be encountered in proposed muon colliders.

Table 1: Straw-Man Muon Collider Parameters

E (TeV)	2	5	50
L (km)	62	36	36
r (m)	3.3	0.8	0.08
Collider depth (m)	300	100	100

The neutrinos exiting a muon collider will not only have a narrow conical shape, but will also have an extent that is quite long. The long, narrow plume of neutrinos will produce secondary muons and hadronic showers at a significant distance from the muon collider. This distance will be greater than tens of kilometers for TeV muon energies.

NEUTRINO INTERACTION MODEL

Neutrinos can interact directly with tissue or with intervening matter to produce charged particles that result in a biological detriment. The radiation environment is complex and simulations (e.g., Monte Carlo methods) can be used to model the dynamics of the neutrino interaction including the energy and angular dependence of each particle (e.g., v_e, $\bar{v}_e$, v_μ, $\bar{v}_\mu$, v_τ, $\bar{v}_\tau$, e, μ, τ, and hadrons) involved in the interaction. Performing a neutrino simulation is too dependent on specific accelerator characteristics and will not add to the health physics presentation. Rather than performing a Monte Carlo simulation, we follow the analytical approach of Cossairt et al. (1997) and King (1999b) to quantify the neutrino effective dose. This approach is acceptable in view of the current uncertainties in muon collider technology and the nature of the neutrino interaction for both charged current (CC) and neutral current (NC) weak processes (King, 1999c).

Following King (1999c), the dominant interaction of TeV-scale neutrinos is deep inelastic scattering with nucleons that include CC and NC components. In the NC process, the neutrino is scattered by a nucleon (N) and loses energy with the production of hadrons (X) through a $\nu+N\rightarrow\nu+X$ reaction. This NC reaction contributes about 25 percent of the total cross section. This NC process can be interpreted as elastic scattering off one of the quarks (q) inside the nucleon through the exchange of a virtual Z^0 boson $(\nu+q\rightarrow\nu+q)$

CC scattering is similar to NC scattering except that the neutrino is converted into its corresponding charged lepton (l). This includes reactions such as $\nu+N\rightarrow l^-+X$ and $\bar{\nu}+N\rightarrow l^++X$ where l is an electron/muon for electron/muon neutrinos. At the quark level, a charged W boson is exchanged with a quark to produce another quark (q') whose charge differs by one unit through processes such as $\nu+q\rightarrow l^-+q'$ and $\bar{\nu}+q'\rightarrow l^++q$. The final state quarks produce hadrons on a nuclear distance scale that contribute to the effective dose. The CC and NC scattering processes are included in the Process A –D descriptions noted in previous discussion.

NEUTRINO EFFECTIVE DOSE

A muon collider provides a platform for colliding beams of muons (m⁻) and antimuons (m⁺) (Geer, 2010). The collider may involve a pair of linear accelerators with intersecting beams or a storage ring that circulates the muons and antimuons in opposite directions prior to colliding the two beams. The accelerator facility energy is usually expressed as the sum of the muon and antimuon energies. For example, a 100 TeV accelerator consists of a 50 TeV muon beam and a 50 TeV antimuon beam. Since muon colliders produce large muon currents, neutrinos will be copiously produced from the decay of both muons and antimuons (See Eqs. 1 and 2).

Neutrino effective dose calculations are performed for two potential muon collider configurations. The first configuration utilizes the intersection of the beams of two muon linear colliders. The linear collider effective dose model incorporates an explicit representation of the neutrino cross section and evaluates the effective dose assuming specific values for the muon energy, number of muon decays per year, and accelerator operational characteristics (e.g., accelerator gradient or the increase in muon energy per unit accelerator length). The operational parameter approach is more familiar to high-energy physicists, but it serves to illustrate the sensitivity of the neutrino effective dose to the key muon collider's operating parameters.

The second configuration is a circular muon collider. The neutrino effective dose for the circular muon collider involves an integral over energy of the differential fluence and fluence to dose conversion factor. This approach is more familiar to health physicists, but much of the muon collider's operating parameters are absorbed into other parameters and are not explicitly apparent. Using both approaches yields not only the desired neutrino effective dose, but also illustrates the sensitivity of the effective dose to a number of accelerator parameters and operational assumptions

BOUNDING NEUTRINO EFFECTIVE DOSE LINEAR MUON COLLIDER

The bounding neutrino effective dose from a linear muon collider is derived following King (1999b) and is based on the effective dose from a straight section (ss) of a circular muon collider. This derivation incorporates a limiting condition from a circular accelerator with a number of straight sections as part of the facility. Parameters unique to the circular collider such as the ring circumference and straight section length appear in intermediate equations, but cancel in the final effective dose result. In the linear muon collider, the muon beam is assumed to be well-collimated.

In a linear muon collider, the total neutrino effective dose (H) is defined in terms of an effective dose contribution $\delta H(E)$ received in each energy interval E to E + dE as the muons accelerate to the beam energy E_o:

$$H = \int_0^{E_o} dE\, \delta H(E)$$

(6)

The effective dose contribution $\delta H(E)$ is written as (King, 1999b):

$$\delta H(E) = H' \frac{1}{f_{ss}} \frac{df(E)}{dE}$$

(7)

where $\frac{df(E)}{dE} dE$ is the fraction of muons that decay via Eqs. 1 and 2 in the energy interval E to E + dE, which may be written as:

$$\frac{df(E)}{dE} = \frac{1}{\gamma \beta c \tau g}$$

(8)

where

$$\gamma = \frac{E_o}{mc^2}$$

(9)

In Eq. 8, $\beta = v / c$, τ is the muon mean lifetime (2.2 x 10^{-6} s), and g is the accelerator gradient (dE/dl). The other parameters appearing in Eq. 7 include fss (the ratio of the straight section length to the ring circumference) and H' (the effective dose that is applicable as the muon energy reaches the TeV energy range), where

$$f_{ss} = \frac{l_{ss}}{C}$$

(10)

In Eq. 10, C is the ring circumference:

$$C = \frac{2\pi E_o}{0.3\overline{B}}$$

(11)

In Eqs. 9 – 11, v is the muon velocity, l_{ss} is the straight section length, E_o is the muon energy, B is the ring's average magnetic induction, and N is the number of muon decays in a year.

In the narrow beam approximation, the effective dose is independent of distance (L) for L < 5 E_o (King, 1999b) where L is expressed in km and E_o in TeV. Using this approximation,

$$H' = K' N l_{ss} \overline{B} E X$$

(12)

where K' is a constant that depends on the units used to express the various quantities appearing in Eq. 12, and X = X(E) is the cross section factor defined in subsequent discussion.

Combining these results leads to the annual neutrino effective dose (H) in mSv/y:

$$H = \frac{NK}{g} \int_0^{E_o} E X(E) dE$$

(13)

where K = 6.7 x 10^{-21} mSv-GeV / m-TeV^2 if g is expressed in GeV/m, N is expressed in muon decays per year, E is the muon energy in TeV, and the cross section factor is dimensionless (Bevelacqua, 2004).

In deriving the linear muon collider effective dose relationship, a number of assumptions were made (Bevelacqua, 2004). These assumptions are explicitly listed to ensure the reader clearly understands the basis for Eq. 13. The relevant assumptions include applicability of the narrow beam approximation. The individual receiving the effective dose is assumed to be: (1) uniformly irradiated, (2) within the footprint of the neutrino beam, (3) within the footprint of the hadronic particle shower that results from the neutrino interactions, and (4) irradiated by only one of the linear muon accelerators whose energy is one-half the total linear muon collider energy. Given the TeV muon energies and the earth shielding present, charged particle equilibrium exists and Process D dominates the neutrino effective dose. In addition, the muon beam is well-collimated, the neutrino effective dose calculation assumes a 100% occupancy factor, and the neutrino effective dose is an annual average based on the number of muon decays in a year.

The cross section factor is a parameterization of the neutrino cross section (See Table 2) in terms of a logarithmic energy interpolation (Quigg, 1997). The numerical factors in the Table 2 expressions (1.453, 1.323, 1.029, 0.512, and 0.175) are the total summed neutrino-nucleon and antineutrino-nucleon cross sections divided by energy at neutrino energies of 0.1, 1, 10, 100, and 1000 TeV, respectively, given in units of 10^{-38} cm^2/GeV. As an approximation, the muon energies in Table 2 are set equal to the corresponding neutrino energies. Following Quigg (1997), the cross section factor is a dimensionless number and is normalized such that $X(E = 0.1 \text{ TeV}) = 1.0$.

Table 2: Cross Section Factor X(E) as a Function of Muon Energy

Muon Energy Range (TeV)	X(E)
E < 1	$(-1.453 \, \alpha + 1.323 \, (\alpha + 1)) / 1.453$
1 < E < 10	$(1.323 \, (1-\alpha) + 1.029 \, \alpha) / 1.453$
10 < E < 100	$(1.029 \, (2-\alpha) + 0.512 \, (\alpha-1)) / 1.453$
100 < E < 1,000	$(0.512 \, (3-\alpha) + 0.175 \, (\alpha-2)) / 1.453$
E > 1,000	$(0.175/1.453) \, 3^{3-\alpha}$
$\alpha = \log_{10}(E)$ where E is the muon energy expressed in TeV.	

Eq. 13 may be approximated by replacing the energy-weighted integral of X(E) by its value at $E = E_o /2$. This choice is acceptable given the energy dependence of the cross section and the associated uncertainties in the collider design parameters. With this selection, the annual neutrino effective dose (mSv/y) becomes:

$$H = \frac{KN}{2g} X(E_o / 2) \, E_o^2$$

$$(14)$$

As a practical example (Zimmerman, 1999), consider a 1,000 TeV muon linear accelerator assuming $E_o = 500$ TeV (i.e., two, 500 TeV linear muon accelerators) and $N = 6.4 \times 10^{18}$ muon decays per year. Using these values in Eq. 14 with a g = 1 GeV/m value leads to an annual effective neutrino dose of 1.4 Sv/y, which is a significant value that cannot be ignored. Health physicists at a linear muon collider will need to contend with large neutrino effective doses within and outside the facility. Table 3 provides expected annual neutrino effective doses for a variety of accelerator energies using the same N and g values noted above and the narrow beam approximation.

Table 3: Annual Neutrino Effective Doses for a Linear Muon Collider Using the Narrow Beam Approximation.

Accelerator Facility Energy (TeV)	Muon Beam Energy (TeV)	H (mSv/y)
0.1	0.05	5.7×10^{-5}
1	0.5	5.2×10^{-3}
10	5	0.45
100	50	30
500	250	440
1,000	500	1.4×10^3
5,000	2,500	1.5×10^4
10,000	5,000	4.2×10^4
50,000	25,000	4.8×10^5

The values of Table 3 suggest that the annual effective dose limit for occupational exposures of 20 mSv/y and the annual effective dose limit to the public (1 mSv/y) can be exceeded by TeV energy muon accelerators (ICRP 103, 2007). The values in Table 3 also exceed the emergency effective dose limit of 250 mSv set for the Fukushima Daiichi accident that is based on ICRP 60 (1991).

A TeV - PeV scale muon collider will also challenge the acute lethal radiation dose ($LD_{50, 30}$) of about 4 Gy (Bevelacqua 2010a). Although the feasibility of TeV - PeV scale machines remains to be determined, the significant radiation hazards associated with their operation merits careful attention to the effects of neutrino effective doses at offsite locations.

Selecting an accelerator location will be an issue for TeV energy muon linear colliders due to public radiation concerns arising from neutrino interactions. Given these radiation concerns, a muon collider location may be restricted to low population or geographically isolated areas to minimize the public neutrino effective dose

Bounding Neutrino Effective dose – Circular muon Collider

The bounding neutrino effective dose for a circular muon collider could be obtained using the methodology of the previous section. However, a number of operational assumptions including the ring circumference and average magnetic induction would be required. Instead, we use an alternative approach to illustrate the various methods than can be utilized to determine the neutrino effective dose as a function of distance. To accomplish this, consider the energy distribution or differential fluence $dN_i(E_i)/dE_i$ where N_i is the number of neutrinos of generation i per unit area, E_i is the neutrino energy, and i = 1, 2, and 3 for the three neutrino generations. The neutrino effective dose H can be determined once the neutrino fluence to effective dose conversion factor $C(E_i)$ is known.

Cossairt et al. (1997) provide an approach for treating the neutrinos and their antiparticles in the first two generations. In view of the limited data, Cossairt et al. (1997) did not consider the generation 3 neutrinos, but these neutrinos become more important as the accelerator energy increases.

One of the initial goals of a muon accelerator will be the development of a pure muon neutrino beam to investigate the magnitude of the neutrino mass. Focusing on the muon neutrino is also warranted because Cossairt et al. (1997) provides a muon neutrino fluence to effective dose conversion factor. Following Cossairt et al. (1997) and Silari & Vincke (2002), we limit the subsequent discussion to muon neutrinos that result from muon decays (Eq. 1) in a circular muon collider and drop the subscript i:

$$H = \int_0^{E_o} \frac{dN(E)}{dE} C(E) dE$$

(15)

where E_o is the energy of the primary muons before decay.

Silari & Vincke (2002) provide a differential fluence value in the laboratory system that is averaged over all neutrino production angles. They also assume the accelerator's shielding is thick enough to attenuate the primary muon beam, and that it is thicker than the range of all secondary radiation. Accordingly, the neutrino radiation is in equilibrium with its secondary radiation.

Using the equilibrium condition and averaging over all production angles, provides the following differential fluence relationship for the neutrino radiation from a circular muon collider (Silari & Vincke, 2002):

$$\frac{dN(E)}{dE} = \frac{2}{E_o}\left(1 - \frac{E}{E_o}\right)\Phi$$

(16)

where N(E) is the number of neutrinos per unit area, E is the neutrino energy, Eo is the energy of the primary muons before decay, and Φ is the integral neutrino fluence (total number of neutrinos per unit area) following the muon decays.

For secondary particle equilibrium, the fluence to effective dose conversion factor relationship of Cossairt et al. (1997) is used:

$$C(E)=KE^2 \tag{17}$$

Eq. 17 was derived for the neutrino energy range of 0.5 GeV to 10 TeV. In deriving the muon neutrino effective dose to fluence conversion factor of Eq. 17, Cossairt et al. (1997) did not consider the effects of the third lepton generation.

In Eq. 17, $K = 10^{-15}$ $\mu Sv\text{-}cm^2/GeV^2$. In view of the trend in the neutrino data (Particle Data Group, 2010; Quigg, 1997), Eq. 17 is used at energies beyond those considered by Cossairt et al. (1997). This is reasonable because increasing energy and increasing number of secondary shower particles (hadrons) is the main reason for the rising fluence to effective dose conversion factor with increasing neutrino energy for the equilibrium (shielded neutrino) case or process D described earlier. It is also reasonable because the neutrino attenuation length (λ) decreases with increasing energy of the primary neutrinos. Although TeV energy units are used in the final result, GeV units are used in the derivation of the neutrino effective dose to facilitate comparison with Silari & Vincke (2002) and Johnson et al. (1998). Prior to developing the neutrino effective dose relationship for a circular muon collider, the neutrino attenuation length is briefly examined.

The neutrino attenuation length is written in terms of the neutrino interaction cross section σ_v:

$$\lambda = \frac{A}{\rho N_A \sigma_v} = \frac{1}{N \sigma_v} \tag{18}$$

where A and ρ are the atomic number and density of the shielding medium, NA is Avogadro's number, N is the number density of atoms of the shielding medium per unit volume, and σ_v is on the order of 10^{-35} cm^2 (E / 1 TeV) (Johnson et al. ,1998) where the neutrino energy is expressed in TeV.

These results permit the neutrino attenuation length to be written as (Johnson et al. ,1998):

$$\lambda = 0.5x10^6 km \left(\frac{1 TeV}{E} \right) \left(\frac{3g / cm^3}{\rho} \right) \tag{19}$$

Since the neutrino attenuation length is very long, the neutrino fluence is very weakly attenuated while traversing a shield. Therefore, shielding is not an effective dose reduction tool for neutrinos.

The effective dose arising from an energy independent neutrino fluence spectrum is accomplished by performing the integration of Eq. 15 using Eqs. 16 and 17:

$$H = \int_0^{E_o} \frac{2}{E_o}\left(1-\frac{E}{E_o}\right)\Phi\left(KE^2\right)dE = \frac{K}{6}E_o^2\Phi$$

(20)

where H is the annual neutrino effective dose in $\dashv$ Sv and Φ is the total number of neutrinos per unit area that is assumed to be independent of energy (Johnson et al. ,1998). The neutrino fluence Φ is the total number of neutrinos traversing a surface behind the shielding. The surface is governed by the divergence of the neutrino beam and the distance r from the neutrino source. The neutrino's half-divergence angle (θ) is:

$$\theta = \frac{mc^2}{E} = \frac{1}{\gamma} \approx \frac{1}{10E_o}$$

(21)

where mc^2 is the muon rest mass in MeV, E is the muon energy, θ is the opening half-angle or characteristic angle of the decay cone expressed in radians, and Eo is the energy of the primary muon beam in GeV.

The neutrino fluence Φ at a given distance r from the muon decay point is just the number of neutrinos N per unit area:

$$\Phi = \frac{N}{\pi(\theta r)^2}$$

(22)

Combining Eqs. 20 - 22 and using the numerical value for K yields a compact form for the annual neutrino effective dose from a circular muon collider:

$$H = \frac{10^{-15} E_o^2}{6}\frac{N}{\pi(\theta r)^2} = \frac{10^{-15} E_o^2}{6}\frac{N}{\pi}\frac{(10E_o)^2}{r^2} = \frac{10^{-13} E_o^4 N}{6\pi r^2}\frac{\mu Sv-cm^2}{GeV^4}$$

(23)

The circular muon collider neutrino effective dose of Eq. 23 has a very strong dependence on the neutrino energy.

Eq. 23 provides the neutrino effective dose assuming all muons decay at the same point. Recognizing that the muons can decay at all storage ring locations with equal probability provides a more physical description of the effective dose. For facilities such as the European Laboratory for Particle Physics (CERN), the neutrino effective dose may to be calculated as an integral over the length of the return arm (l) (Silari & Vincke, 2002) of the storage ring pointing toward the surface from d to d + l, where d is the thickness of material traversed by the neutrino beam between the end of the return arm and the surface of the earth along the direction of the return arm. The quantity d may also be described as the approximate minimum thickness of earth needed to absorb the circulating muons if beam misdirection or total beam loss occurs (i.e., the beam exits the facility). Recognizing that the muons may decay at any location along the return arm, leads to the neutrino effective dose:

$$H = \frac{10^{-13} E_o^4}{6\pi} \int_d^{d+l} \frac{N}{l} \frac{dr}{r^2} = \frac{10^{-13} E_o^4 N}{6\pi l} \left(\frac{1}{d} - \frac{1}{d+l} \right) \frac{\mu S\upsilon - cm^2}{GeV^4}$$

(24)

Silari & Vincke (2002) provides parameters for the planned muon facility at CERN. For a 50 GeV muon energy in the storage ring, $N = 1021$ muons per year decaying in the ring, a return arm length pointing toward the surface ($l = 6.0 \times 10^4$ cm), and a 100 m thickness of material (d) traversed by the neutrino beam between the end of the return arm and the surface, a surface neutrino effective dose of 47 mSv/yr is predicted. Since the planned CERN design has 3 return arms, the effective dose rate at the end of one of the arms would be about 16 mSv/y (47 mSv/3). Increasing muon energy will lead to higher muon effective dose rates, additional muon shielding requirements, and will force the collider deeper underground (See Table 4, derived from Silari & Vincke, (2002).

Table 4: Geometrical Parameters for Representative Cases of Circular Muon Colliders

Muon Energy (TeV)	d (m)	L (km)	φ (mrad)	θ (μrad)
1	100	36	5.6	106
2	100	36	5.6	53
5	200	51	8	21
10	500	80.5	12.5	11

These results suggest that the circular muon collider be installed underground to shield the muon beam in the event the beam becomes misdirected. This required shielding is determined by the muon energy loss (Silari & Vincke, 2002):

$$\frac{dE}{dx} = 0.6 \frac{TeV}{km} \left(\frac{\rho}{3g/cm^3} \right)$$

(25)

When compared to muons, neutrinos have a much smaller interaction cross section. The earth shielding that completely attenuates the muons will have a negligible effect on the neutrinos. Accordingly, the neutrinos will produce a nontrivial annual effective dose at the earth's surface where the beam emerges. In order to evaluate the magnitude of this neutrino effective dose, assume the earth is a sphere, and a horizontal, circular muon collider is situated a depth d below the earth's surface. The neutrino beam exit point from the earth will be at a horizontal distance L given by Silari & Vincke (2002):

$$L = \sqrt{2dR - d^2} \approx \sqrt{2dR} \approx 36\,km \sqrt{\frac{d}{100\,m}}$$

(26)

where R = 6400 km is the earth's radius. Table 4 provides representative values of d and L. In addition to d and L, a number of other relevant parameters associated with the circular collider of Eq. 26 are summarized in Table 4. In Table 4, φ is the half-angle subtended by the horizontal accelerator beam with respect to the earth's center before it exits the earth:

$$Sin\,\varphi = L\,/\,R$$

(27)

The functional form of Eq. 24 suggests that the calculation of neutrino effective dose from a circular muon collider is dependent of the assumed physical configuration and beam characteristics. An estimate of the neutrino effective dose for a circular muon collider can be made using Eq. 23. For comparison with Eq. 14, Eq. 23 is rewritten in terms of TeV and mSv units:

$$H = \frac{10^{-4}\,E_o^4\,N}{6\pi r^2}\,\frac{mSv - cm^2}{TeV^4}$$

(28)

where N is the number of muon decays per year, Eo is the muon energy in TeV, r is the distance from the point of muon decay in cm, and H is the annual neutrino effective dose in mSv. For consistency with the linear muon collider assumptions, 6.4x1018 muon decays per year are assumed in subsequent calculations. Given the TeV muon energies and the earth shielding present, charged particle equilibrium is assumed to exist. Moreover, the neutrino beam is limited to muon neutrinos only.

The muon neutrino effective dose to fluence conversion factor is assumed to be valid at energies beyond those utilized in Cossairt et al. (1997). Given the TeV muon energies, Process D of Cossairt et al. (1997) will dominate the neutrino effective dose.

In deriving the circular muon collider effective dose relationship, a number of assumptions were made. First, the neutrino effective dose calculation assumes a 100% occupancy factor, and is an annual average based on the number of muon decays in a year. Second, the muon beam is well-collimated. In addition, the irradiated individual is (1) assumed to be within the footprint of the neutrino beam and the hadronic particle shower that results from the neutrino interactions, (2) irradiated by only one of the muon beam's decay neutrinos whose energy is one-half the total circular muon collider energy, and (3) uniformly irradiated by the neutrino and hadronic radiation types.

Table 5 summarizes the results of neutrino effective dose values as a function of distance from the muon decay location (r) for a circular muon collider. Since the facility energy is the sum of the muon and antimuon energies, a 100 TeV accelerator consists of a 50 TeV muon beam and a 50 TeV antimuon beam.

The long, thin conical radiation plumes present a radiation challenge well beyond the facility boundary. For example, a 25 TeV circular muon collider produces a neutrino effective dose of 37 mSv/y at a distance of 1500 km from the facility. Although the neutrino effective dose plume will only have a radius of 12 m at 1500 km, it presents a radiation challenge for muon collider health physicists and management. The effective dose values summarized in Table 5 have the potential to impart lethal doses to small areas. The large effective dose values and their control must be addressed in facility design and licensing.

The importance of properly characterizing offsite public effective doses is illustrated by the Fukushima Daiichi Nuclear Power Station (FDNPS) accident in Japan (Butler; 2011a, 2011b). These doses focused attention on inadequacies in the FDNPS design and licensing bases. Offsite effective doses and their profile must be carefully and credibly addressed in muon collider design and licensing evaluations.

Table 5: Annual Neutrino Effective Doses for a Circular Muon Collider.

Accelerator Energy (TeV)[a]	H (mSv/y) at the Specified Distance (r) from the Accelerator				
	5 km	25 km	100 km	1500 km	2500 km
0.1	8.5×10^{-4}	3.4×10^{-5}	2.1×10^{-6}	9.4×10^{-9}	3.4×10^{-9}
2	140	5.4	0.34	1.5×10^{-3}	5.4×10^{-4}
25	3.3×10^{6}	1.3×10^{5}	8.3×10^{3}	37	13
100	8.5×10^{8}	3.4×10^{7}	2.1×10^{6}	9.4×10^{3}	3.4×10^{5}
500	5.3×10^{11}	2.1×10^{10}	1.3×10^{9}	5.9×10^{6}	2.1×10^{6}
1000	8.5×10^{12}	3.4×10^{11}	2.1×10^{10}	9.4×10^{7}	3.4×10^{7}

Physics and cost parameters associated with 0.1, 3, 10, and 100 TeV circular muon colliders (King 1999a) are summarized in Table 6. Given current levels of technology, the collider cost will present a funding challenge as TeV muon energies are reached. In addition to funding issues, the control of radiation from the muon beams and neutrino plumes must be addressed. The feasibility of higher energy colliders will necessarily depend on technological development as well as financial support of scientific agencies.

Table 6: Circular Muon Collider Physics and Cost Parameters

Accelerator Energy (TeV)	0.1	3	10	100
Circumference (km)	0.35	6	15	100
Average Magnetic Field (T)	3.0	5.2	7.0	10.5
Cost	Feasible	Challenging	Challenging	Problematic

As the collider energy increases, muon shielding requirements dictate a subsurface facility. The impact of locating the muon collider deeper underground with increasing accelerator energy can also be investigated. Using Eq. 28 and the data summarized in Table 4, permit the calculation of the neutrino effective dose upon its exit from the earth's surface. If the same beam properties are assumed as for the linear muon collider (i.e., $N = 6.4 \times 10^{18}$ muon decays per year) and $r = L$ (Table 4), then the magnitude and size of the resultant radiation plumes derived from Eq. 28 are summarized in Table 7.

Table 7: Neutrino Effective Dose Characteristics for a Circular Muon Collider

Muon Energy (TeV) [a]	d (m) [b]	L (Horizontal Distance at the Earth's Surface) (km) [c]	Beam Radius at the Earth's Surface (m) [d]	H at the Earth's Surface (mSv/y)
1	100	36	3.6	2.6
2	100	36	1.8	42
5	200	51	1.0	820
10	500	80.5	0.8	5.2×10^3
50	500	80.5	0.16	3.3×10^6
100	500	80.5	0.081	5.2×10^7
500	500	80.5	0.016	3.3×10^{10}
1000	500	80.5	0.0081	5.2×10^{11}

[a] The accelerator energy is twice the muon energy.

[b] Accelerator depth below the surface of the earth.

[c] Horizontal exit point distance from the surface of the earth.

[d] The half-divergence angle is determined from Eq. 5.

Although the effective dose results at the earth's surface are significant, they occur over a relatively small area. The results also assume a 100% occupancy factor for this small area, which is not likely. The magnitude of the neutrino effective dose merits significant attention and emphasis on radiation monitoring and control. For example, a 500 TeV muon beam would deliver an acute absorbed dose rate of about 1 Gy/s to a 3.2 cm diameter circle. This absorbed dose rate is sufficient to deliver a biological detriment to the body within seconds (Bevelacqua, 2010a).

Dose management controls will be similar to those enacted for direct beam exposures at conventional accelerators. Interlocks associated with beam misalignment are effective in limiting the probability that the beam is directed toward an unanticipated direction. However, additional methods to control the offsite neutrino dose must be developed because lethal exposures can occur in a very short time even though the areas involved are small. Subjecting the public to potentially lethal effective doses represents unique facility licensing challenges that must be addressed in facility safety analyses. Public perception and stakeholder involvement will be key elements in licensing TeV-PeV scale muon colliders. The need for public involvement in licensing and regulatory discussions becomes particularly important when high effective doses could result from facility operations.

OFFSITE EFFECTIVE DOSE CONSIDERATIONS FOR MUON COLLIDERS

TeV energy neutrinos do not behave according to conventional operational health physics experience at power reactors and contemporary accelerator facilities. As noted previously, neutrinos are electrically uncharged and only interact through the weak interaction. Their small, but non-zero, interaction cross section creates a unique situation in terms of the behavior of the neutrino effective dose, particularly in terms of the shape and energy dependence of their radiation profile. These properties will lead to a modification of conventional health physics dose reduction concepts when applied to planned muon colliders.

Basic radiation protection principles suggest that the effective dose at a given location is reduced if the exposure time is minimized, the distance from the source is increased, or shielding is added between the source and the point of interest (Bevelacqua, 2009, 2010a). These principles must be modified at a TeV energy muon collider. The time principle is still valid for muons and neutrinos. The neutrino and muon effective doses are reduced by decreasing the exposure time.

The distance principle is ineffective when neutrinos are involved. Since neutrinos interact very weakly, relatively long distances are not effective in significantly reducing the neutrino effective dose. In fact, the neutrino beam remains a hazard for hundreds of kilometers. However, distance will still be effective for reducing the muon effective dose.

Unlike other radiation types, shielding neutrinos increases the effective dose. The magnitude of the particle showers produced by neutrino interactions is governed by the quantity of shielding material between the neutrino beam and the point of interest. However, shielding muons is an effective dose reduction measure. From the standpoint of TeV energy neutrino radiation, a linear muon collider has a number of advantages over circular muon colliders.

Firstly, the radiation is confined to two narrow beams that can be oriented to minimize the interaction of the neutrinos. A simple dose reduction technique orients the linear accelerators at an angle such that the neutrino beams exit the accelerator above the ground. This configuration minimizes the residual neutrino interactions with the earth and man-made structures. Secondly, the spent muons can be removed from the beam following collisions or interactions before they decay into highenergy neutrinos.

OTHER RADIATION PROTECTION ISSUES

A number of radiation protection issues associated with TeV energy muon colliders will challenge accelerator health physicists. The issues related to large neutrino effective dose values and effective neutrino dosimetry were previously noted. Before construction of a muon collider, thorough studies will be performed to define the accelerator's radiation footprint. These studies will: (1) define muon collider shielding requirements; (2) assess induced activity within the facility and the environment (e.g., air, water, and soil), including the extent of groundwater activation; (3) assess radiation streaming through facility penetrations (e.g., ventilation ducts and access points); (4) assess various accident scenarios such as loss of power or beam misdirection; and (5) assess the various pathways for liquid and airborne releases of radioactive material. Facility waste generation and decommissioning are other areas that will require evaluation.

In addition to the aforementioned radiation protection issues, the TeV energy neutrino beam will create new issues. Radiation protection concerns unique to muon colliders have been reported by Autin et al. (1999), Bevelacqua (2004), Johnson et al. (1998), Mokhov & Cossairt (1998), and Mokhov et al. (2000). These authors suggest that above about 1.5 TeV, the neutrino induced secondary radiation will pose a significant hazard even at distances on the order of tens to hundreds of kilometers. The neutrino radiation hazard presents both a physical as well as political challenge (King, 1999a).

These issues also complicate the process for locating a suitable site for a TeV energy muon collider. There are a number of potential solutions to reduce the neutrino effective dose associated with a muon collider. These include using radiation boundaries or fenced-off areas to denote areas with elevated effective dose values. Building the collider on elevated ground or at an isolated area would also minimize human exposure. Effective dose reduction measures are also available for specific muon collider configurations.

In a linear muon collider operating at the higher TeV energies, dose reduction is achieved by locating the interaction region above the earth's surface. In a circular muon collider, dose reduction is achieved by minimizing the straight sections in the ring, burying the collider deep underground to increase the distance before the neutrino beam exits the ground, and orienting the collider ring to take advantage of natural topographical features.

Orders of magnitude reductions in the neutrino effective dose are required for the muon colliders noted in this chapter (See Tables 3, 5, and 7) to meet current regulations for public exposures (ICRP, 2007). Some of the possible effective dose reduction solutions may be difficult to implement for the TeV energy muon colliders. The most feasible options for locating and operating the highest TeV energy muon collider are to either use (1) an isolated location where no one is exposed to the neutrino radiation before it exits into the atmosphere as a result of the earth's curvature, or (2) a linear muon collider constructed such that the individual muon beams collide in air well above the earth's surface.

For Option 1, the accelerator could either be constructed at an elevated location or at an isolated area. The area will need to be large, perhaps having a site boundary with a diameter greater than 100 km (King, 1999a). This size requirement restricts the available locations, and would normally require that the facility have access to the resources of an existing accelerator facility such as CERN or Fermilab. Alternatively, the facility could be located in an isolated area and scientific personnel relocated to that area with the establishment of a self-sufficient site. The final decision regarding facility location will involve funding and political considerations that are part of new facility development, licensing, and construction.

Option 2 would be technically feasible, and could be located at a smaller site. However, design considerations for both Options 1 and 2 would need to address a number of potential radiation issues associated with accelerator operation (Bevelacqua, 2008, 2009, and 2010a) that could lead to significant, unanticipated radiation levels in controlled as well as uncontrolled areas. Radiation protection issues include beam alignment errors, design errors, unauthorized changes, activation sources, and control of miscellaneous radiation sources (Bevelacqua, 2008, 2009, 2010a). These operational issues require close control because they have the potential to produce large and unanticipated effective dose values.

Beam alignment errors could direct the beam in unanticipated directions. Given the long range of the muon effective dose profile, these errors could have a significant impact on licensing and accident analysis. Beam alignment errors are caused by a variety of factors including power failures, maintenance errors, and magnet failures. Both human errors and mechanical failures lead to beam alignment issues. Changes in the beam energy or beam current, that exceed the authorized operating envelope, lead to elevated fluence rates, the creation of unanticipated particles, or the creation of particles with higher energy than anticipated. Changes to beam parameters must be carefully evaluated for their impact on the radiation environment of the facility.

The control of secondary radiation sources, radio-frequency equipment, high-voltage power supplies, and other experimental equipment merits special attention. These sources of radiation are more difficult to control than the primary or scattered accelerator radiation because health physicists may not be aware of their existence, the experimenters may not be aware of the hazard, or the radiation source is at least partially masked by the accelerator's radiation output. These miscellaneous radiation sources will include x-rays as well as other types of radiation.

OVERVIEW OF THE NEUTRINO EFFECTIVE DOSE AT A TAU COLLIDER

A third generation tau collider has not been evaluated. In order to provide an estimate of the effective dose consequences of a tau collider, a modification of the muon collider methodology is utilized. The decay characteristics of a tau are considerably more complex than muon decay. The muon essentially decays with a branching ratio of 100 % into a lepton and neutrinos via Eq. 1. For example, tau decays involve 119 decay modes with specified branching fractions with six modes accounting for 90% of the decays (Particle Data Group 2010). The dominant tau decay mode is:

$$\tau^- \to \pi^- + \pi^0 + \nu_\tau \ (25.51\%)$$

$$(29)$$

However, the negative pion dominantly decays into a muon and antimuon neutrino, and the neutral pion decays primarily into photons.

$$\tau^- \to \left(\mu^- + \bar{\nu}_\mu\right) + \left(\gamma + \gamma\right) + \nu_\tau$$

$$(30)$$

Subsequently, the muon decays following Eq. 1. Eq. 30 then yields:

$$\tau^- \to \left(e^- + \nu_\mu + \bar{\nu}_e + \bar{\nu}_\mu\right) + \left(\gamma + \gamma\right) + \nu_\tau$$

$$(31)$$

The net result of the decay is that multiple neutrinos are produced from the tau and subsequent decay of particles. The factor ξ described in subsequent discussion incorporates the effects of the multiple tau decay modes and their effects on the neutrino effective dose.

Subsequent discussion assumes no annihilation of particles and antiparticles in the beam produced by the tau decay products. In addition, the narrow beam approximation is assumed.

The neutrino dose from tau decays is determined by comparing the number of neutrinos emitted from an equal number of tau and muon decays. ξ defines the ratio of the number of neutrinos contributing to the tau collider to muon collider effective doses:

$$\xi = \frac{\displaystyle\sum_{i=1}^{N} Y_i \sum_{j=1}^{3} \left(a n_i \left(v_j \right) + b n_i \left(\bar{v}_j \right) \right)}{a n \left(v_\mu \right) + b n \left(\bar{v}_e \right)}$$

(32)

In the numerator of Eq. 32, i labels the various decay modes of the tau, N is the number of tau decay modes, Yi is the branching fraction of the ith tau decay mode, $n_i(v_j)$ is the number of generation j neutrinos emitted from decay mode i, and $n_i\left(\bar{v}_j\right)$ is the number of generation j antineutrinos emitted from decay mode i. In the denominator of Eq. (32), $n\left(\bar{v}_e\right)$ is the number of muon neutrinos emitted in a muon decay, and n(ve) is the number of antielectron neutrinos emitted in a muon decay. The j sum counts the three neutrino generations, and a and b are the cross-section factors of King (1999a) for neutrinos and antineutrinos which are 1.0 and 0.5, respectively.

The ratio of tau neutrino to muon neutrino effective doses is obtained by utilizing the value of ξ and the calculated ratio of tau and muon neutrino cross-sections (β) (Jeong & Reno, 2010). The discussion is applicable to circular and linear muon and tau colliders. For equivalent accelerator operating conditions (e.g., beam energy and number of beam particle decays) and receptor conditions (e.g., distance and ambient conditions), the ratio of neutrino effective doses from a tau collider and muon collider is given by:

$$\frac{H_{\tau^-}(E)}{H_{\mu^-}(E)} = \xi \beta(E)$$

(33)

The results of calculations utilizing Eq. 33 are summarized in Table 8.

Table 8: Ratio of Tau and Muon Collider Neutrino Effective Doses

Beam Energy (TeV)	Effective Dose Ratio
0.01	0.39
0.1	1.75
1.0	2.16
10.	2.23

The tau collider neutrino effective doses are generally larger than those encountered in a muon collider, and the tau dose profile is also larger. The larger tau profile is demonstrated by considering Eqs. 3 and 4 for equivalent tau and muon collider configurations:

$$\frac{r(\tau^-)}{r(\mu^-)} = \frac{m_{\tau^-}}{m_{\mu^-}} = \frac{1777\,MeV}{105.7\,MeV} = 16.8$$

(34)

Using Eq. 34 and the Table 7 results for circular tau collider conditions, the neutrino effective dose profile radius at the earth's surface is 60.5, 30.2, 16.8, and 13.4 m for 1, 2, 5, and 10 TeV beams. These affected areas and associated effective doses suggest that the tau collider is a more significant radiation hazard than the muon collider. Therefore, larger effective doses and affected areas are anticipated during tau collider operations.

An improved calculation of the neutrino effective dose from a tau collider requires a better specification of neutrino properties. For example, previous calculations were based on the Standard Model assumption that neutrinos have zero mass. Neutrino masses can be calculated assuming the alternative gauge group $SU(2)_L \otimes SU(2)_R \otimes U(1)$ instead of the Standard Model $SU(2)_L \otimes U(1)$. This gauge group leads to a neutrino generation i mass:

$$m_i = \frac{M_i^2}{g\, m_{W_R}}$$

(35)

where M_i is the generation i lepton mass ($(e, \mu, and\ \tau)$, W_R is the right-handed W boson mass (≥ 300 GeV), and g is a coupling constant with a value of 0.585 (Mohapatra & Senjanović, 1980). Using these values in Eq. 35 leads to electron, muon, and tau neutrino upper bound masses of 1.5 eV, 64 keV, and 18 MeV, respectively. These masses affect the input values used to calculate the neutrino effective dose in Eqs. 14 and 23. As an alternative, better crosssection data and dose conversion factors would refine the neutrino effective dose.

CONCLUSIONS

Neutrino radiation will be a health physics issue and design constraint for muon colliders, particularly at TeV energies. TeV energy muon colliders will require careful site selection and the neutrino effective dose may dictate that these machines be constructed in isolated areas. With the operation of TeV energy muon colliders, the neutrino effective dose can no longer be neglected. Neutrino detection, neutrino dosimetry, and the determination of the neutrino effective dose will no longer be academic exercises, but will become operational health physics concerns. Keeping public and occupational neutrino effective doses below regulatory limits will require careful and consistent application of dose reduction methods.

When compared to muon colliders, initial scooping calculations for tau colliders suggest that higher effective doses and affected areas will result from their operation. Although, the tau collider calculations are initial estimates, they suggest that significant radiation challenges are also presented by these machines.

REFERENCES

1. Autin, B; Blondel, A. & Ellis, J. (1999). Prospective Study of Muon Storage Rings at CERN, CERN 99-02, European Laboratory for Particle Physics, Geneva, Switzerland Bevelacqua, J. (2004). Muon Colliders and Neutrino Dose Equivalents: ALARA Challenges for the 21st Century, Radiation Protection Management, Vol.21, No. 4, pp. 8-30.

2. Bevelacqua, J. (2008). Health Physics in the 21st Century, Wiley-VCH, ISBN 9783527408221, Weinheim, Germany

3. Bevelacqua, J. (2009). Contemporary Health Physics: Problems and Solutions (Second Edition), ISBN 9783527408245, Weinheim, Germany

4. Bevelacqua, J. (2010a). Basic Health Physics: Problems and Solutions (Second Edition), ISBN 9783527408238, Weinheim, Germany

5. Bevelacqua, J. (2010b). Standard Model of Particle Physics-A Health Physics Perspective, Health Physics, Vol.99, No.5, pp. 613-623

6. Butler, D. (2011a). Radioactivity Spreads in Japan, Nature, Vol.471, No.7340, pp. 555-556

7. Butler, D. (2011b). Fukushima Health Risks Scrutinized, Nature, Vol.472, No.7341, pp. 13-14

8. Cottingham, W. & Greenwood, D. (2007). An Introduction to the Standard Model of Particle Physics (Second Edition), Cambridge University Press, ISBN 9780521852494, Cambridge, UK

9. Cossairt, J.; Grossman, N. & Marshall, E. (1996). Neutrino Radiation Hazards: A Paper Tiger, Fermilab-Conf-96/324, Accessed on July 11, 2011, Available from: <http://lss.fnal.gov/archive/1996/conf/Conf-96-324.pdf>

10. Cossairt, J.; Grossman, N. & Marshall, E. (1997). Assessment of Dose Equivalent due to Neutrinos, Health Physics, Vol.73, No.6, 894-898.

11. Cossairt, J. & Marshall, E. (1997). Comment on "Biological Effects of Stellar Collapse Neutrinos, Physical Review Letters, Vol.78, No.7, pp.1394.

12. Collar, J. (1996). Biological Effects of Stellar Collapse Neutrinos, Physical Review Letters, Vol.76, No.6, pp. 999-1002

13. Geer, S. (2010). From Neutrino Factory to Muon Collider, FERMILAB-CONF-10-024-APC, Accessed on July 14, 2011, Available from: < http:// arxiv.org/abs/1006.0923> Griffiths, D. (2008). Introduction to Elementary Particle Physics (Second Edition), Wiley-VCH, ISBN 9783527406012, Weinheim, Germany

14. ICRP Report No. 60. (1991). 1990 Recommendations of the International Commission on Radiological Protection, Elsevier, Amsterdam ICRP Report No. 107. (2007). The 2007 Recommendations of the International Commission on Radiological Protection, Elsevier, Amsterdam

15. Jeong, Y. & Reno, M. (2010). Tau neutrino and antineutrino cross sections, Accessed on July 12, 2011, Available from: <http://arxiv.org/ PS_cache/arxiv/pdf/1007/1007.1966v1.pdf>

16. Johnson, C.; Rolandi, G. & Silari, M. (1998). Radiological Hazard due to Neutrinos from a Muon Collider, Internal Report CERN/TIS-RP/IR/98, European Laboratory for Particle Physics (CERN), Geneva, Switzerland (1998). Accessed on July 12, 2011, Available from: <http://www.physics. princeton.edu/mumu/johnson/neutrino.pdf>

17. King, B. (1999a). Studies for Muon Collider Parameters at Center-of-Mass Energies of 10 TeV and 100 TeV, Brookhaven National Laboratory, Accessed on July 25, 2011, Available from: <http://arxiv.org/PS_cache/ physics/pdf/9908/9908018v1.pdf >

18. King, B. (1999b). Neutrino Radiation Challenges and Proposed Solutions for Many-TeV Muon Colliders, Proc. HEMC'99 Workshop – Studies on Colliders and Collider Physics at the Highest Energies: Muon Colliders at 10 TeV to 100 TeV, Montauk, NY,

19. September 1999 Accessed on July 25, 2011, Available from: <http:// nfmccdocdb.fnal.gov/cgibin/RetrieveFile?docid=119&version=1&filena me=muc0119.ps.gz>

20. King, B. (1999c). Neutrino Physics at Muon Colliders, Brookhaven National Laboratory, Accessed on July 19, 2011, Available from: <http:// arxiv.org/PS_cache/hepex/pdf/9907/9907035v1.pdf>

21. Kuno, Y. (2009). Project X Workshop Summary, Muon Collider Physics Workshop, Fermi National Laboratory, Batavia, IL, 10 -12 November, 2009, Available from: <https://indico.fnal.gov/getFile.py/access?contrib Id=78&sessionId=0&resId=0& materialId=slides&confId=2855>

22. Mohapatra, R. & Senjanović, G. (1980). Neutrino Mass and Spontaneous Parity Nonconservation, Physical Review Letters, Vol.44, No.14, pp. 912-915

23. Mokhov, N. & Cossairt, J. (1998). Radiation Studies at Fermilab, Proceedings of the Fourth Workshop on Simulating Accelerator Radiation Environments (SARE4), Knoxville, TN, 14 – 16 September, 1998, Accessed on July 25, 2011, Available from: <http://lss.fnal.gov/archive/1998/conf/Conf-98-384.pdf >

24. Mokhov, N.; Striganov, S. & van Ginneken, A. (2000). Muons and Neutrinos at High Energy Accelerators, FERMILAB-Conf-00/182, Accessed on July 11, 2011, Available from: <http://lss.fnal.gov/archive/2000/conf/Conf-00-182.pdf>

25. NCRP Report No. 144. (2003). Radiation Protection for Particle Accelerator Facilities, National Council on Radiation Protection and Measurements, ISBN 0929600770, Bethesda, MD Particle Data Group. (2010). Review of Particle Properties, Journal of Physics G, Vol.37, No.7A, pp. 1-1422.

26. Quigg, C. (1997). Neutrino Interaction Cross Sections, FERMILAB-Conf-97/158-T, Accessed on July 11, 2011, Available from: < http://lss.fnal.gov/archive/1997/conf/Conf-97-158-T.pdf>

27. Silari, M. & Vincke, H. (2002). Neutrino Radiation Hazard at the Planned CERN Neutrino Factory, Technical Note TIS-RP/TN/2002-01, Accessed on July 12, 2011, Available from: <http://slap.web.cern.ch/slap/NuFact/NuFact/nf105.pdf>

28. Zimmerman, F. (1999). Final Focus Challenges for Muon Colliders at Highest Energies, Proc. HEMC'99 Workshop – Studies on Colliders and Collider Physics at the Highest Energies: Muon Colliders at 10 TeV to 100 TeV, Montauk, NY, September 1999, Accessed on July 25, 2011, Available from: <//cdsweb.cern.ch/record/420774/files/sl-1999-077.ps.gz>

29. Zisman, M. (2011). R&D Toward a Neutrino Factory and Muon Collider, LBNL-4494E, awrence Berkeley National Laboratory, Berkeley, CA, Accessed on July 25, 2011, Available from: <http://escholarship.org/uc/item/43p7z0v1>

Chapter 9

SOFT GLUON RESUMMATION FOR ASSOCIATED $t\bar{t}H$ PRODUCTION AT THE LHC

Anna Kulesza,[a] Leszek Motyka,[b] Tomasz Stebel[b] and Vincent Theeuwes[a]

[a]Institute for Theoretical Physics, WWU Münster, Münster, D-48149 Germany

[b] Institute of Physics, Jagellonian University, S. Lojasiewicza 11, Kraków, 30-348 Poland

ABSTRACT

We perform resummation of soft gluon corrections to the total cross section for the process $pp \to t\bar{t}H$. The resummation is carried out at next-to-leading-logarithmic (NLL) accuracy using the Mellin space technique, extending its application to the class of $2 \to 3$ processes. We present an analytical result for the soft anomalous dimension for a hadronic production of two coloured massive particles in association with a colour singlet. We discuss the impact of resummation on the numerical prediction for the associated Higgs boson production with top quarks at the LHC.

INTRODUCTION

Establishing the properties of the Higgs boson discovered at the LHC in 2012 [1, 2], in particular its couplings to the Standard Model (SM) particles, is one of the main tasks of the current LHC run. Since the SM Higgs boson couples to fermions proportionally to their masses, the top-Higgs Yukawa coupling is expected to be especially sensitive to the underlying physics. A direct way to probe the strength of the coupling without making any assumptions regarding its nature is provided by the measurement of Higgs production rates in the $pp \to t\bar{t}H$ process. Although the production cross section is low and the collision energy and the luminosity available so far have not been sufficient enough to measure a Higgs signal in Run 1 [3–7], such a measurement in Run 2 is eagerly awaited. Correspondingly, precision predictions for the $pp \to t\bar{t}H$ production process are of great importance and a lot of effort has been invested in the recent years to improve the theoretical accuracy. The next-to-leading-order (NLO) QCD, i.e. $\mathcal{O}(\alpha_s^3\alpha)$ predictions are already known for some time [8–13] and have been newly recalculated and matched to parton showers in [14– 17]. As of late, the mixed QCD-weak corrections [18] and QCD-EW corrections [19, 20] of $\mathcal{O}(\alpha_s^2\alpha^2)$ are also available.

Furthermore, the NLO QCD corrections to the hadronic ttH⁻ production with top and antitop quarks decaying into bottom quarks and leptons have been recently obtained [21]. Concurrently, new methods for a better measurement of the process have been proposed e.g. in [22] or in [23]. In general, for the LHC collision energies of Run 2, the NLO QCD corrections are ~ 20%, whereas the size of the (electro)weak correction is more than ten times smaller. The scale uncertainty of the NLO QCD corrections is estimated to be ~ 10% [8–13, 24]. While matching fixed-order predictions to parton showers pursued recently by many groups in such frameworks as aMC@NLO [14, 15, 25], POWHEG BOX [16, 17, 26] or SHERPA [27] allows for a more accurate description of final state characteristics, it does not change the predictions for the overall production rates. An improvement in the accuracy with which these rates are known can only be achieved by calculating higher order corrections. However, calculations of the next-to-next-to-leadingorder corrections are currently technically out of reach. It is nevertheless interesting to ask the question what is the size and the effect of certain classes of corrections of higher than NLO accuracy. In particular, we focus here on taking into account contributions from soft gluon emission to all orders in perturbation theory. The traditional (Mellinspace) resummation formalism which is applied in this type of calculations has been very well developed and copiously employed for description of the 2 → 2 type processes at the Born level. The universality of resummation concepts warrants their applications to scattering processes with many partons in the final state, as shown in a general analytical treatment developed for arbitrary number of partons [28–30]. Recently, the soft gluon resummation technique in the soft collinear effective theory (SCET) framework was applied to $pp \rightarrow t\bar{t}W^{\pm}$ [31]. So far, however, no calculations in the traditional resummation framework for processes involving 2 → 3 scattering at the Born level have been performed.

In this paper we take the first step in this direction by developing the Mellin-space threshold resummation formalism at the next-to-leading-logarithmic (NLL) accuracy for the case of 2 → 3 processes with two coloured massive particles in the final state. We then apply this formalism in order to estimate the impact of soft gluon corrections on the predictions for the total ttH⁻ production rate. In this particular case, the threshold region is reached when the square of the partonic center-of-mass (c.o.m.) energy, $\sqrt{\hat{s}}$, approaches M

$= 2m_t + m_H$, where mt is the top quark mass and mH is the Higgs boson mass. In the threshold region, the cross section receives enhancement in the form of logarithmic corrections in $\beta = \sqrt{1 - M^2/\hat{s}}$. The quantity β measures the distance from absolute production threshold and can be related to the maximal velocity of the tt⁻ system. Additionally, in the threshold region the virtual QCD corrections are also enhanced due to Coulomb-type interactions between the two final state top quarks which become large when the top quark velocity in the $t\bar{t}$ c.o.m. frame $\beta_{kl} \to 0$ with $\beta_{kl} = \sqrt{1 - 4m_t^2/\hat{s}_{kl}}$ $\hat{s}_{kl} = (p_t + p_{\bar{t}})^2$. However, the contributions to the total cross section from the threshold region are strongly suppressed by the β^4 factor originating from the massive three particle phase space. Nevertheless, one expects that the threshold corrections can still have a non-negligible impact on the predictions.

The associated production of a Higgs boson with a tt⁻ pair involves four colored partons at the Born level and as such is characterized by a non-trivial color flow. The color structure influences the contributions from wide-angle soft gluon emissions which have to be included at the NLL accuracy. The evolution of the color exchange at NLL is governed by the one-loop soft anomalous dimension [28, 32–36]. Starting from four colored partons in the process, the soft anomalous dimension is a matrix and is known for heavyquark [32, 33, 37], dijet [34–36] and supersymmetric particle production [38–40, 43], as well as for the general case of 2 → n QCD processes [28–30]. Here we adopt the calculations of the soft anomalous dimension for the case of 2 → 3 processes with two colored massive particles in the final state.

RESUMMATION FOR 2 → 3 PROCESSES WITH TWO MASSIVE COLORED PARTICLES IN THE FINAL STATE

The resummation of soft gluon corrections to the total cross section $\sigma_{pp \to t\bar{t}H}$ is performed in Mellin space, where the Mellin moments are taken w.r.t. the variable $\rho = M^2/S$. At the partonic level, the Mellin moments for the process $ij \to klB$, where i, j denote massless colored partons, k, l two massive quarks and B a massive color-singlet particle, is given by

$$\hat{\sigma}_{ij \to klB,N}(m_k, m_l, m_B, \mu_F^2, \mu_R^2) = \int_0^1 d\hat{\rho}\, \hat{\rho}^{N-1} \hat{\sigma}_{ij \to klB}(\hat{\rho}, m_k, m_l, m_B, \mu_F^2, \mu_R^2)$$

(1)

with $\hat{\rho} = 1 - \beta^2$.

At LO, the $t\bar{t}H$ production receives contributions from the qq⁻ and gg channels. We analyze the colour structure of the underlying processes in the s-channel color bases, $\{c_I^q\}$ and $\{c_I^g\}$, with $c_1^q = \delta^{\alpha_i\alpha_j}\delta^{\alpha_k\alpha_l}$, $c_8^q = T_{\alpha_i\alpha_j}^a T_{\alpha_k\alpha_l}^a$, $c_1^g = \delta^{a_i a_j}\delta^{\alpha_k\alpha_l}$, $c_{8S}^g = T_{\alpha_l\alpha_k}^b d^{ba_i a_j}$,

$\mathscr{C}_{8A}^g = iT^b_{\alpha_l \alpha_k} f^{b\alpha_i \alpha_j}$. In this basis the soft anomalous dimension matrix becomes diagonal in the production threshold limit [32, 33] and the NLL resummed cross section in the N-space has the form [32, 33, 37]

$$\hat{\sigma}^{(res)}_{ij \to klB, N} = \sum_I \hat{\sigma}^{(0)}_{ij \to klB, I, N} \, C_{ij \to klB, I} \, \Delta^i_{N+1} \Delta^j_{N+1} \Delta^{(int)}_{ij \to klB, I, N+1},$$

(2)

where we suppress explicit dependence on the scales. The index I in eq. (2) distinguishes between contributions from different colour channels. The colour-channel-dependent contributions to the LO partonic cross sections in Mellin-moment space are denoted by $\hat{\sigma}^{(0)}_{ij \to klB, I, N}$. The radiative factors Δ^i_N describe the effect of the soft gluon radiation collinear to the initial state partons and are universal. Large-angle soft gluon emission is accounted for by the factors $\Delta^{(int)}_{ij \to klB, I, N}$ which depend on the partonic process under consideration and the colour configuration of the participating particles. The expressions for the radiative factors in the $\overline{MS}$ factorisation scheme read (see e.g. [37])

$$\ln \Delta^i_N = \int_0^1 dz \frac{z^{N-1} - 1}{1 - z} \int_{\mu_F^2}^{M^2(1-z)^2} \frac{dq^2}{q^2} A_i(\alpha_s(q^2)),$$

$$\ln \Delta^{(int)}_{ij \to klB, I, N} = \int_0^1 dz \frac{z^{N-1} - 1}{1 - z} D_{ij \to klB, I}(\alpha_s(M^2(1-z)^2)).$$

(3)

The coefficients A_i, $D_{ij \to klB, I}$ are power series in the coupling constant α_s,

$$A_i = \left(\frac{\alpha_s}{\pi}\right) A_i^{(1)} + \left(\frac{\alpha_s}{\pi}\right)^2 A_i^{(2)} + \dots, \qquad D_{ij \to klB, I} = \left(\frac{\alpha_s}{\pi}\right) D^{(1)}_{ij \to klB, I} + \dots$$

(4)

The universal LL and NLL coefficients $A_i^{(1)}$, $A_i^{(2)}$ are well known [44, 45] and given by

$A_i^{(1)} = C_i$, $A_i^{(2)} = \frac{1}{2} C_i \left(\left(\frac{67}{18} - \frac{\pi^2}{6}\right) C_A - \frac{5}{9} n_f \right)$ with $C_g = C_A = 3$, and $C_q = C_F = 4/3$.

The NLL coefficients $D_{ij \to klB, I}$ are obtained by taking the threshold limit $\hat{s} \to M^2 = (m_k + m_l + m_B)^2$ of the gauge-invariant soft anomalous dimension matrices $\Gamma^{ij \to klB}$. In this limit $\Gamma^{ij \to klB} = \frac{\alpha_s}{\pi} \text{diag}(\gamma_1^{ij}, \dots)$ and $D_{ij \to klB, I} = 2\text{Re}(\gamma_I^{ij})$. The calculations of $\Gamma^{ij \to klB}$ apply the methods developed in the heavy quark pair-production [32, 33] to the process at hand, taking into account $2 \to 3$ kinematics, and yield

$$\Gamma^{q\bar{q}\to klB} = \frac{\alpha_s}{\pi} \begin{bmatrix} -C_F(L_{\beta,kl}+1) & \frac{C_F}{C_A}\Omega_3 \\ 2\Omega_3 & \frac{1}{2}[(C_A - 2C_F)(L_{\beta,kl}+1) + C_A\Lambda_3 + (8C_F - 3C_A)\Omega_3] \end{bmatrix}, \tag{5}$$

$$\Gamma^{gg\to klB} = \frac{\alpha_s}{\pi} \begin{bmatrix} \Gamma^{gg}_{11} & 0 & \Omega_3 \\ 0 & \Gamma^{gg}_{22} & \frac{N_c}{2}\Omega_3 \\ 2\Omega_3 & \frac{N_c^2-4}{2N_c}\Omega_3 & \Gamma^{gg}_{33} \end{bmatrix}, \tag{6}$$

with

$$\Gamma^{gg}_{11} = -C_F(L_{\beta,kl}+1),$$

$$\Gamma^{gg}_{22} = \Gamma^{gg}_{33} = \frac{1}{2}((C_A - 2C_F)(L_{\beta,kl}+1) + C_A\Lambda_3),$$

where

$$\Lambda_3 = (T_1(m_k) + T_2(m_l) + U_1(m_l) + U_2(m_k))/2,$$

$$\Omega_3 = (T_1(m_k) + T_2(m_l) - U_1(m_l) - U_2(m_k))/2,$$

and

$$L_{\beta,kl} = \frac{\kappa^2 + \beta_{kl}^2}{2\kappa\beta_{kl}} \left(\log\left(\frac{\kappa - \beta_{kl}}{\kappa + \beta_{kl}}\right) + i\pi \right), \tag{7}$$

$$T_i(m) = \frac{1}{2}\left(\ln((m^2 - t_i)^2/(m^2\hat{s})) - 1 + i\pi\right), \tag{8}$$

$$U_i(m) = \frac{1}{2}\left(\ln((m^2 - u_i)^2/(m^2\hat{s})) - 1 + i\pi\right), \tag{9}$$

$$\kappa = \sqrt{1 - (m_k - m_l)^2/s_{kl}}, \qquad s_{kl} = (p_k + p_l)^2, \tag{10}$$

$$t_1 = (p_i - p_k)^2, \qquad t_2 = (p_j - p_l)^2, \qquad u_1 = (p_i - p_l)^2, \qquad u_2 = (p_j - p_k)^2. \tag{11}$$

Eqs. (5), (6) reproduce the known results for heavy quark-antiquark (squark-antisquark) pair- production soft anomalous dimension [32, 33, 38–40] in the limit pB $\to$ 0. Also, our result for $\Gamma^{q\bar{q}\to klB}$ agrees with the result obtained in the SCET framework in [41, 42]. It can be also explicitly seen that in the limit $\hat{s} \to (2m_t + m_H)^2$ the non-diagonal elements vanish and the diagonal elements give $D_{q\bar{q}\to klB,I} = \{0, -N_c\}$, $D_{gg\to klB,I} = \{0, -N_c, -N_c\}$,, which are the same coefficients as for the heavy-quark pair production $D_{ij\to kl}$. This confirms a simple physical intuition that the properties of the soft emission in the absolute threshold limit are only driven by the colour structure of the subprocesses and do not depend on the their kinematics.

For completness we display the explicit NLL expressions for the resummed factors in the Mellin space, which were used in our numerical implementation:

$$\ln \Delta_N^i \overset{\text{NLL}}{=} g_i^{(1)} \left(b_0 \, \alpha_s(\mu_R^2) \ln N \right) \ln N \; + \; g_i^{(2)} \left(b_0 \, \alpha_s(\mu_R^2) \ln N, M^2, \mu_R^2, \mu_F^2 \right) , \tag{12}$$

$$\ln \Delta_{ij \to kl\, B, I, N}^{(\text{int})} \overset{\text{NLL}}{=} h_{ij \to kl\, B, I}^{(2)} \left(b_0 \, \alpha_s(\mu_R^2) \ln N \right) \tag{13}$$

with

$$g_i^{(1)}(\lambda) = \frac{A_i^{(1)}}{2\pi b_0 \lambda} \left[2\lambda + (1 - 2\lambda) \ln(1 - 2\lambda) \right] , \tag{14}$$

$$
\begin{aligned}
g_i^{(2)}(\lambda, M^2, \mu_R^2, \mu_F^2) = & -\frac{A_i^{(1)} \gamma_E}{\pi b_0} \ln(1 - 2\lambda) \\
& + \frac{A_i^{(1)} b_1}{2\pi b_0^3} \left[2\lambda + \ln(1 - 2\lambda) + \frac{1}{2} \ln^2(1 - 2\lambda) \right] \\
& - \frac{A_i^{(2)}}{2\pi^2 b_0^2} \left[2\lambda + \ln(1 - 2\lambda) \right] \\
& - \frac{A_i^{(1)}}{2\pi b_0} \left[2\lambda + \ln(1 - 2\lambda) \right] \ln \left(\frac{\mu_R^2}{M^2} \right) \\
& + \frac{A_i^{(1)}}{2\pi b_0} 2\lambda \ln \left(\frac{\mu_F^2}{M^2} \right) ,
\end{aligned}
\tag{15}
$$

$$h_{ij \to kl\, B, I}^{(2)}(\lambda) = \frac{\ln(1 - 2\lambda)}{2\pi b_0} D_{ij \to kl\, B, I} , \tag{16}$$

where b_0 and b_1 are the first two coefficients of the QCD β-function

$$b_0 = \frac{11 C_A - 4 T_R n_f}{12\pi} , \qquad b_1 = \frac{17 C_A^2 - 10 C_A T_R n_f - 6 C_F T_R n_f}{24\pi^2} . \tag{17}$$

The coefficients

$$C_{ij \to kl B, I} = 1 + \frac{\alpha_s}{\pi} C_{ij \to kl B, I}^{(1)} + \cdots \tag{20}$$

contain all non-logarithmic contributions to the NLO cross section taken in the threshold limit. More specifically, these consist of Coulomb corrections, N-independent hard contributions from virtual corrections and N-independent non-logarithmic contributions from soft emissions. Although formally the coefficients $C_{ij \to klB,I}$ begin to contribute at NNLL accuracy, in our numerical studies of the $pp \to t\bar{t}H$ process we consider both the case of $C_{ij \to klB,8}^{(1,\mathrm{Coul})} = (C_F - C_A/2)\pi^2/(2\beta_{kl})$, i.e. with the first-order corrections to the coefficients neglected, as well as the case with these corrections included. In the latter case we treat the Coulomb corrections and the hard contributions additively, i.e.

$$C_{ij \to klB,I}^{(1)} = C_{ij \to klB,I}^{(1,\mathrm{hard})} + C_{ij \to klB,I}^{(1,\mathrm{Coul})}.$$

For k, l denoting massive quarks the Coulomb corrections are $C_{ij \to klB,1}^{(1,\mathrm{Coul})} = C_F \pi^2/(2\beta_{kl})$ and $C_{ij \to klB,8}^{(1,\mathrm{Coul})} = (C_F - C_A/2)\pi^2/(2\beta_{kl})$. The additive treatment is consistent with NLL resummation and matching to NLO. We note that in general Coulomb corrections can also be resummed [46–50]. A combined resummation of Coulomb and soft corrections is, however, beyond the scope of this paper.

THEORETICAL PREDICTIONS FOR THE $pp \to t\bar{t}H$ PROCESS AT NLO+NLL ACCURACY

The resummation-improved NLO+NLL cross sections for the $pp \to t\bar{t}H$ process are obtained through matching the NLL resummed expressions with the full NLO cross sections

$$\hat{\sigma}_{h_1 h_2 \to kl}^{(\mathrm{NLO+NLL})}(\rho, \mu_F^2, \mu_R^2) = \hat{\sigma}_{h_1 h_2 \to klB}^{(\mathrm{NLO})}(\rho, \mu_F^2, \mu_R^2) + \hat{\sigma}_{h_1 h_2 \to klB}^{(\mathrm{res-exp})}(\rho, \mu_F^2, \mu_R^2)$$

$$\text{with} \quad \hat{\sigma}_{h_1 h_2 \to klB}^{(\mathrm{res-exp})} = \sum_{i,j} \int_C \frac{dN}{2\pi i} \rho^{-N} f_{i/h_1}^{(N+1)}(\mu_F^2) f_{j/h_2}^{(N+1)}(\mu_F^2)$$

$$\times \left[\hat{\sigma}_{ij \to klB,N}^{(\mathrm{res})}(\mu_F^2, \mu_R^2) - \hat{\sigma}_{ij \to klB,N}^{(\mathrm{res})}(\mu_F^2, \mu_R^2) \Big|_{(\mathrm{NLO})} \right], \quad (21)$$

where $\hat{\sigma}_{ij \to klB,N}^{(\mathrm{res})}$ is given in eq. (2) and $\hat{\sigma}_{ij \to klB,N}^{(\mathrm{res})}|_{(\mathrm{NLO})}$ represents its perturbative expansion truncated at NLO. The moments of the parton distribution functions (pdf) $f_{i/h}(x, \mu_F^2)$ are defined in the standard way $f_{i/h}^{(N)}(\mu_F^2) \equiv \int_0^1 dx\, x^{N-1} f_{i/h}(x, \mu_F^2)$. The inverse Mellin transform (21) is evaluated numerically using a contour C in the complex-N space according to the "Minimal Prescription" method developed in ref. [51].

As mentioned in the previous section, the calculation of first-order contributions to the coefficients $C_{ij \to t\bar{t}H,I}$ requires knowledge of the NLO real corrections in the threshold limit as well as virtual corrections. In our calculations we follow the methodology of [52, 53], where the case of two massive coloured particle in the final state was considered. We have explicitly checked that adding a massive colour singlet particle in the final state does not introduce any extra terms dependent on the mass of the added particle. Thus the N-space results for the pair-production process of two massive coloured particles are also applicable in our $2 \rightarrow 3$ case. This way, the problem of calculating the $C^{(1)}_{ij \rightarrow t\bar{t}H,I}$ coefficients reduces to calculation of virtual corrections to the process. We extract them numerically using the publicly available POWHEG implementation of the $t\bar{t}H$ process [17], based on the calculations developed in [10–13]. The results were then cross-checked using the standalone MadLoop implementation in aMC@NLO [14]. Since the $q\bar{q}$ channel receives only colour-octet contributions, the extracted value contributing to $C^{(1,\text{hard})}_{q\bar{q} \rightarrow t\bar{t}H,8}$ is exact. In the gg channel, however, both the singlet and octet production modes contribute. The implementation of the virtual corrections to $gg \rightarrow t\bar{t}H$ in POWHEG and in aMC@NLO does not allow for their separate extraction in each colour channel. Instead, we extract the value which contributes to the coefficient $\bar{C}^{(1,\text{hard})}_{gg \rightarrow t\bar{t}H}$ averaged over colour channels and use the same value to further calculate $C^{(1,\text{hard})}_{gg \rightarrow t\bar{t}H,1}$ and $C^{(1,\text{hard})}_{gg \rightarrow t\bar{t}H,8}$. In order to measure the size of the error introduced by this procedure, we then rescale this value by the ratios of the corresponding colour-channel dependent and colour averaged coefficients found for $gg \rightarrow t\bar{t}H$ in [54].

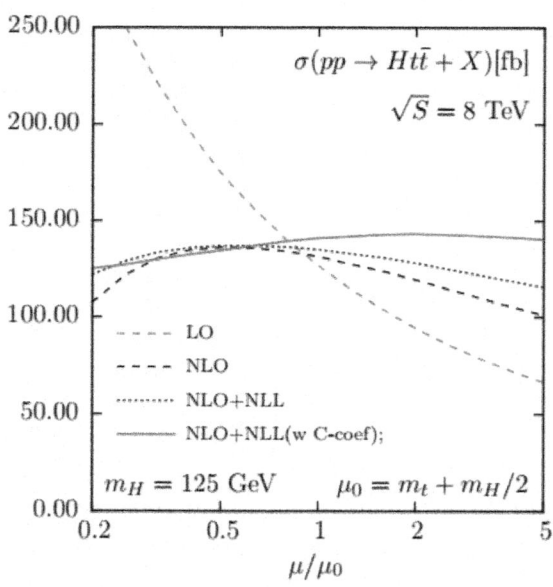

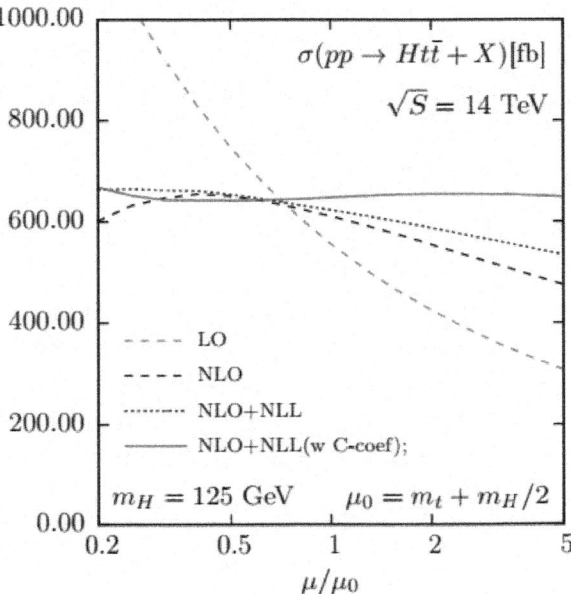

The scale dependence of the $C^{(1)}_{ij\to klB,I}$ can be fully deduced from renormalization group arguments, in the same way as for the full NLO result. We have checked that numerical results obtained with the procedure which we use to extract the values of the coefficients at $\mu_0 = \mu_F = \mu_R$ show the same scale dependence as expected from exact analytical expressions.

Figure 1: Scale dependence of the LO, NLO and NLO+NLL cross sections at $\sqrt{S} = 8$ and $\sqrt{S} = 14$ TeV LHC collision energy. The results are obtained while simultaneously varying μ_F and μ_R, $\mu = \mu_F = \mu_R$.

In our phenomenological analysis we use $m_t = 173$ GeV, $m_H = 125$ GeV and choose the central scale $\mu_{F,0} = \mu_{R,0} = m_t + m_H/2$, in accordance with [24]. The NLO cross section is calculated using the aMC@NLO code [25]. In the implementation of the resummation formula, eq. (2), we numerically take a Mellin transform of the LO cross sections and the $C^{(1)}_{ij\to t\bar{t}H,I}$ coefficient terms which are both calculated in the x space. We perform the current analysis employing MMHT2014 [55] pdfs and use the corresponding values of αs. Beside presenting the full result including non-zero $C^{(1)}_{ij\to t\bar{t}H,I}$ coefficients, we also show the results with $C_{ij\to t\bar{t}H,I} = 1$.

We begin our numerical study by analysing the scale dependence of the resummed total cross section for $pp \to t\bar{t}H$ at $\sqrt{S} = 8$ and 14 TeV varying simultaneously the factorization and renormalization scales, μF and μR. As demonstrated in figure 1, adding the soft gluon corrections stabilizes the dependence on $\mu = \mu_F = \mu_R$ of the NLO+NLL predictions with respect to NLO. As an example, the central values and the scale error at $\sqrt{S} = 8\,\mathrm{TeV}$ changes from $132^{+3.9\%}_{-9.3\%}$ fb at NLO to $141^{+1.4\%}_{-4.2\%}$ fb at NLO+NLL (with $C^{(1)}_{ij \to t\bar{t}H,I}$ coefficients included) and correspondingly, from $613^{+6.2\%}_{-9.4\%}$ fb to $650^{+0.8\%}_{-1.2\%}$ fb at $\sqrt{S} = 14\,\mathrm{TeV}$. It is also clear from figure 1 that the coefficients $C^{(1)}_{ij \to t\bar{t}H}$ strongly impact the predictions, especially at higher scales.

In order to understand these effects better, in figure 2 we analyse the dependence on the factorization and renormalization scale separately for the case study of $\sqrt{S} = 14\,\mathrm{TeV}$. We observe that the weak scale dependence present when the scales are varied simultaneously is a result of the cancellations between renormalization and factorization scale dependencies.

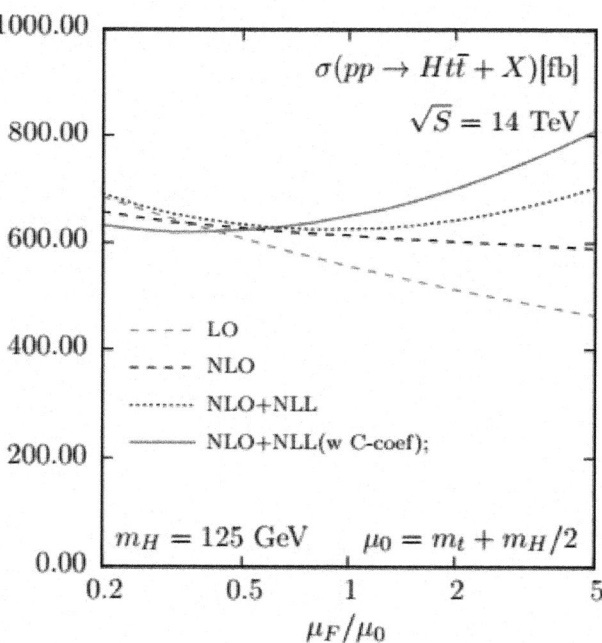

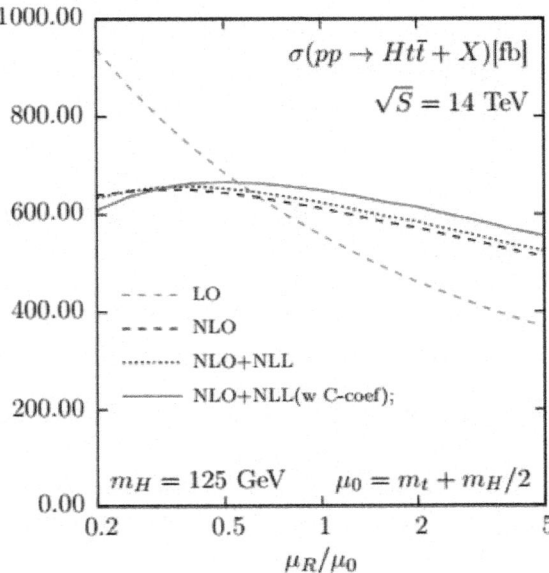

Figure 2: Factorization and renormalization scale dependence of the LO, NLO and NLO+NLL cross sections at $\sqrt{S} = 14\,\text{TeV}$ LHC collision energy. The results are obtained with $\mu_R = \mu_0$ for μ_F variation and and $\mu_F = \mu_0$ for μ_R variation.

A similar effect of the opposite behavior of the total cross section under μ_F and μ_R variations was previously shown for the total cross section for the inclusive Higgs production in the gluon-fusion process [57]. The typical decrease of the cross section with increasing μ_R originates from running of α_s. The behaviour under variation of the factorization scale, on the other hand, is related to the effect of scaling violation of pdfs at probed values of x. In this context, it is interesting to observe that the NLO+NLL predictions in figure 2 show very little μ_F dependence around the central scale, in agreement with expectation of the factorization scale dependence in the resummed exponential and in the pdfs cancelling each other, here up to NLL. The relatively strong dependence on µF of the NLO+NLL predictions with non-zero $C_{ij\to t\bar{t}H,I}^{(1)}$ can be then easily understood: the resummed expression will take into account higher order scale dependent terms which involve both $C_{ij\to t\bar{t}H,I}^{(1)}$ and logarithms of N. These terms do not have their equivalent in the pdf evolution since the pdfs do not carry any process-specific information. Correspondingly, they are not cancelled and can lead to strong effects if the coefficients $C_{ij\to t\bar{t}H,I}^{(1)}$ are numerically substantial. As these terms can only provide a part of the full scale dependence at higher orders, it is to be expected that their impact will be significantly modified when NNLO corrections are known.

Given the arguments above, we choose to estimate the theoretical uncertainty due to scale variation using the 7-point method, where the minimum and maximum values obtained with $(\mu_F/\mu_0, \mu_R/\mu_0) = (0.5, 0.5),(0.5, 1),(1, 0.5),(1, 1),(1, 2),(2, 1),(2, 2)$ are considered. The effect of including NLL corrections is summarized in table 1 for the LHC collision energy of 8, 13 and 14 TeV. The NLO+NLL predictions show a significant reduction of the scale uncertainty, compared to NLO results. The reduction of the positive and negative scale errors amounts to around 20-30% of the NLO error for $\sqrt{S} = 13, 14\,\mathrm{TeV}$ and to around 25–35% for $\sqrt{S} = 8\,\mathrm{TeV}$. This general reduction trend is not sustained for the positive error after including the $C^{(1)}_{ij \to t\bar{t}H,I}$ coefficients. More specifically, the negative error is further slightly reduced, while the positive error is increased. The origin of this increase can be traced back to the substantial dependence on μ_F of the resummed predictions with non-zero $C^{(1)}_{ij \to t\bar{t}H,I}$ coefficients, manifesting itself at larger scales.

Table 1: NLO+NLL and NLO total cross sections for $pp \to t\bar{t}H$ for various LHC collision energies. The error ranges given together with the NLO and NLO+NLL results indicate the scale uncertainty.

$\sqrt{S}$ [TeV]	NLO [fb]	NLO+NLL		NLO+NLL with C		pdf error
		Value [fb]	K-factor	Value [fb]	K-factor	
8	$132^{+3.9\%}_{-9.3\%}$	$135^{+3.0\%}_{-5.9\%}$	1.03	$141^{+7.7\%}_{-4.6\%}$	1.07	$+3.0\%$ -2.7%
13	$506^{+5.9\%}_{-9.4\%}$	$516^{+4.6\%}_{-6.5\%}$	1.02	$537^{+8.2\%}_{-5.5\%}$	1.06	$+2.3\%$ -2.3%
14	$613^{+6.2\%}_{-9.4\%}$	$625^{+4.6\%}_{-6.7\%}$	1.02	$650^{+7.9\%}_{-5.7\%}$	1.06	$+2.3\%$ -2.2%

However, even after the redistribution of the error between the positive and negative parts, the overall size of the scale error, corresponding to the size of the error bar, is reduced after resummation by around 7% at 8 TeV and 10 (13)% at 13 (14) TeV with respect to the NLO uncertainties. The scale error of the predictions is still a few times larger than the pdf error, cf. table For simplicity, the pdf error shown in table 1 is calculated for the NLO predictions, however adding the soft gluon correction can only minimally influence the value of the pdf error.[1]

As expected on the basis of large phase-space suppression in the threshold regime, the predictions for total cross section at NLO+NLL are only moderately increased by 2–3% w.r.t. the full NLO result. Introducing the coefficients

$C^{(1)}_{ij \to t\bar{t}H,I}$ leads to an increase in the K-factor of up to 6–7%, indicating the importance of constant terms in the threshold limit. Since the impact of soft corrections is bigger for processes taking place closer to threshold the K-factor gets slightly higher for smaller collider energies. We also check the impact of our approximated treatment of keeping parts of $C^{(1,\text{hard})}_{gg \to t\bar{t}H,1}$ and $C^{(1,\text{hard})}_{gg \to t\bar{t}H,8}$ coefficients coming from the virtual corrections equal to the colour channel averaged value, by rescaling at $\mu_F = \mu_R = \mu_0$ the averaged $\bar{C}^{(1,\text{hard})}_{gg \to t\bar{t}H}$ coefficient with ratios $C^{(1,\text{hard})}_{gg \to t\bar{t},I}/\bar{C}^{(1,\text{hard})}_{gg \to t\bar{t}}$ taken from [54]. The procedure is motivated by obvious similarities between the colour structures of the $pp \to t\bar{t}$ and $pp \to t\bar{t}H$ cross sections considered at threshold. We find that such rescaling of the hadronic $t\bar{t}H$ cross section leads to a 3 per mille effect at 14 TeV, or a 5% effect on the correction itself. Therefore we do not expect that the exact knowledge of the $C^{(1)}_{gg \to t\bar{t}H,1}$ and $C^{(1)}_{gg \to t\bar{t}H,8}$' coefficients will have a significant impact on the hadronic NLO+NLL predictions. However, we stress that because of the large phase-space suppression in the threshold regime the resummed results, while systematically taking into account a well-defined class of correction, should not be used to estimate the size of the NNLO total cross section, by e.g. methods of expansion of the resummed exponential.

In more detail, QCD corrections to the $t\bar{t}H$ cross section may be divided into logarithmically enhanced (up to the NLL accuracy) soft gluon corrections and the formally subleading pieces i.e. corrections that enter beyond the NLL accuracy at the absolute threshold limit. A direct numerical analysis of relative importance of the two classes of corrections in the NLO correction shows that the soft gluon logarithms do not dominate the exact NLO correction. In the window of renormalisation and factorisation scales $1/2\mu_0 < \mu_F = \mu_R < 2\mu_0$ the NLL result expanded to the NLO accuracy differs from the NLO cross-section by about 10%, which should be compared to the typical relative magnitude of the exact NLO correction in this scale window of up to 20%. We took into account some of these formally non-leading corrections via the C (1)-coefficient determined in the absolute threshold limit. This approach, however, also does not provide a satisfactory approximation to the exact NLO correction. We conclude that a good approximation of the exact NLO correction requires inclusion of subleading pieces in the NLL expansion beyond the absolute threshold limit. Therefore our results should be viewed as an all-order improvement of a well-defined sub-class of perturbative corrections to the $t\bar{t}H$ cross-section, which, however, omits other possibly important contributions in the full perturbative expansion.

SUMMARY

We have investigated the impact of the soft gluon emission effects on the total cross section for the process $pp \to t\bar{t}H$ at the LHC. The resummation of soft gluon emission has been performed using the Mellin-moment resummation technique at the NLO+NLL accuracy. To the best of our knowledge, this is the first application of this method to a $2 \to 3$ process. Supplementing the NLO predictions with NLL corrections results in moderate modifications of the overall size of the total rates. The size of these modifications, as well as the size of the theoretical error due to scale variation is strongly influenced by the inclusion of the first-order hard matching coefficients into the resummation framework. The overall size of the theoretical scale error becomes smaller after resummation, albeit the reduction is relatively modest when the non-zero first-order hard matching coefficients are considered.

Note added. After the arXiv publication of this paper, ref. [58] appeared. The results of [58] seem to support our conclusion regarding the importance of the corrections from beyond the absolute threshold region.

ACKNOWLEDGMENTS

This work has been supported in part by the DFG grant KU 3103/ Support of the Polish National Science Centre grants no. DEC-2014/13/B/ST2/02486 is gratefully acknowledged. TS acknowledges support in the form of a scholarship of Marian Smoluchowski Research Consortium Matter Energy Future from KNOW funding

REFERENCES

1.   ATLAS collaboration, Observation of a new particle in the search for the Standard Model Higgs boson with the ATLAS detector at the LHC, Phys. Lett. B 716 (2012) 1

2.   CMS collaboration, Observation of a new boson at a mass of 125 GeV with the CMS experiment at the LHC, Phys. Lett. B 716 (2012) 30

3.   ATLAS collaboration, Search for H $\to$ $\gamma\gamma$ produced in association with top quarks and constraints on the Yukawa coupling between the top quark and the Higgs boson using data taken at 7 TeV and 8 TeV with the ATLAS detector, Phys. Lett. B 740 (2015) 222

4. ATLAS collaboration, Search for the Standard Model Higgs boson produced in association with top quarks and decaying into b ̄b in pp collisions at √ s = 8 TeV with the ATLAS detector, Eur. Phys. J. C 75 (2015) 349

5. CMS collaboration, Search for the associated production of the Higgs boson with a top-quark pair, JHEP 09 (2014) 087 Erratum ibid. 10 (2014) 106.

6. CMS collaboration, Search for a Standard Model Higgs Boson Produced in Association with a Top-Quark Pair and Decaying to Bottom Quarks Using a Matrix Element Method, Eur. Phys. J. C 75 (2015) 251

7. ATLAS collaboration, Search for the associated production of the Higgs boson with a top quark pair in multilepton final states with the ATLAS detector, Phys. Lett. B 749 (2015) 519

8. W. Beenakker, S. Dittmaier, M. Kr ̈amer, B. Plumper, M. Spira and P.M. Zerwas, Higgs radiation off top quarks at the Tevatron and the LHC, Phys. Rev. Lett. 87 (2001) 201805

9. W. Beenakker, S. Dittmaier, M. Kr ̈amer, B. Plumper, M. Spira and P.M. Zerwas, NLO QCD corrections to ttH⁻ production in hadron collisions, Nucl. Phys. B 653 (2003) 151

10. L. Reina and S. Dawson, Next-to-leading order results for tth⁻ production at the Tevatron, Phys. Rev. Lett. 87 (2001) 201804

11. L. Reina, S. Dawson and D. Wackeroth, QCD corrections to associated tth⁻ production at the Tevatron, Phys. Rev. D 65 (2002) 053017

12. S. Dawson, L.H. Orr, L. Reina and D. Wackeroth, Associated top quark Higgs boson production at the LHC, Phys. Rev. D 67 (2003) 071503

13. S. Dawson, C. Jackson, L.H. Orr, L. Reina and D. Wackeroth, Associated Higgs production with top quarks at the large hadron collider: NLO QCD corrections, Phys. Rev. D 68 (2003) 034022

14. V. Hirschi, R. Frederix, S. Frixione, M.V. Garzelli, F. Maltoni and R. Pittau, Automation of one-loop QCD corrections, JHEP 05 (2011) 044

15. R. Frederix, S. Frixione, V. Hirschi, F. Maltoni, R. Pittau and P. Torrielli, Scalar and pseudoscalar Higgs production in association with a top-antitop pair, Phys. Lett. B 701 (2011) 427

16. M.V. Garzelli, A. Kardos, C.G. Papadopoulos and Z. Tr ́ocs ́anyi, Standard Model Higgs boson production in association with a top anti-top pair at NLO with parton showering, Europhys. Lett. 96 (2011) 11001

17. H.B. Hartanto, B. Jager, L. Reina and D. Wackeroth, Higgs boson production in association with top quarks in the POWHEG BOX, Phys. Rev. D 91 (2015) 094003

18. S. Frixione, V. Hirschi, D. Pagani, H.-S. Shao and M. Zaro, Weak corrections to Higgs hadroproduction in association with a top-quark pair, JHEP 09 (2014) 065

19. Y. Zhang, W.-G. Ma, R.-Y. Zhang, C. Chen and L. Guo, QCD NLO and EW NLO corrections to ttH⁻ production with top quark decays at hadron collider, Phys. Lett. B 738 (2014) 1

20. S. Frixione, V. Hirschi, D. Pagani, H.-S. Shao and M. Zaro, Electroweak and QCD corrections to top-pair hadroproduction in association with heavy bosons, JHEP 06 (2015) 184

21. A. Denner and R. Feger, NLO QCD corrections to off-shell top-antitop production with leptonic decays in association with a Higgs boson at the LHC, JHEP 11 (2015) 209

22. T. Plehn, G.P. Salam and M. Spannowsky, Fat Jets for a Light Higgs, Phys. Rev. Lett. 104 (2010) 111801

23. P. Artoisenet, P. de Aquino, F. Maltoni and O. Mattelaer, Unravelling tth via the Matrix Element Method, Phys. Rev. Lett. 111 (2013) 091802

24. LHC Higgs Cross Section Working Group, S. Dittmaier et al., Handbook of LHC Higgs Cross Sections: Inclusive Observables, arXiv:1100593

25. J. Alwall et al., The automated computation of tree-level and next-to-leading order differential cross sections and their matching to parton shower simulations, JHEP 07 (2014) 079

26. S. Alioli, P. Nason, C. Oleari and E. Re, A general framework for implementing NLO calculations in shower Monte Carlo programs: the POWHEG BOX, JHEP 06 (2010) 043

27. T. Gleisberg et al., Event generation with SHERPA 1, JHEP 02 (2009) 007

28. R. Bonciani, S. Catani, M.L. Mangano and P. Nason, Sudakov resummation of multiparton QCD cross-sections, Phys. Lett. B 575 (2003) 268

29. S.M. Aybat, L.J. Dixon and G.F. Sterman, The Two-loop anomalous dimension matrix for soft gluon exchange, Phys. Rev. Lett. 97 (2006) 072001

30. S.M. Aybat, L.J. Dixon and G.F. Sterman, The Two-loop soft anomalous dimension matrix and resummation at next-to-next-to leading pole, Phys. Rev. D 74 (2006) 074004

31. H.T. Li, C.S. Li and S.A. Li, Renormalization group improved predictions for ttW⁻ ± production at hadron colliders, Phys. Rev. D 90 (2014) 094009

32. N. Kidonakis and G.F. Sterman, Subleading logarithms in QCD hard scattering, Phys. Lett. B 387 (1996) 867

33. N. Kidonakis and G.F. Sterman, Resummation for QCD hard scattering, Nucl. Phys. B 505 (1997) 321

34. J. Botts and G.F. Sterman, Hard Elastic Scattering in QCD: Leading Behavior, Nucl. Phys. B 325 (1989) 62

35. N. Kidonakis, G. Oderda and G.F. Sterman, Threshold resummation for dijet cross-sections, Nucl. Phys. B 525 (1998) 299

36. N. Kidonakis, G. Oderda and G.F. Sterman, Evolution of color exchange in QCD hard scattering, Nucl. Phys. B 531 (1998) 365

37. R. Bonciani, S. Catani, M.L. Mangano and P. Nason, NLL resummation of the heavy quark hadroproduction cross-section, Nucl. Phys. B 529 (1998) 424

38. A. Kulesza and L. Motyka, Threshold resummation for squark-antisquark and gluino-pair production at the LHC, Phys. Rev. Lett. 102 (2009) 111802

39. A. Kulesza and L. Motyka, Soft gluon effects in supersymmetric particle production at the LHC, Acta Phys. Polon. B 40 (2009) 1957

40. A. Kulesza and L. Motyka, Soft gluon resummation for the production of gluino-gluino and squark-antisquark pairs at the LHC, Phys. Rev. D 80 (2009) 095004

41. A. Ferroglia, M. Neubert, B.D. Pecjak and L.L. Yang, Two-loop divergences of scattering amplitudes with massive partons, Phys. Rev. Lett. 103 (2009) 201601

42. A. Ferroglia, M. Neubert, B.D. Pecjak and L.L. Yang, Two-loop divergences of massive scattering amplitudes in non-abelian gauge theories, JHEP 11 (2009) 062

43. W. Beenakker, S. Brensing, M. Kr̈amer, A. Kulesza, E. Laenen and I. Niessen, Soft-gluon resummation for squark and gluino hadroproduction, JHEP 12 (2009) 041

44. J. Kodaira and L. Trentadue, Summing Soft Emission in QCD, Phys. Lett. B 112 (1982) 66

45. S. Catani, E. D'Emilio and L. Trentadue, The Gluon Form-factor to Higher Orders: Gluon-Gluon Annihilation at Small Qt, Phys. Lett. B 211 (1988) 335

46. V.S. Fadin and V.A. Khoze, Threshold Behavior of Heavy Top Production in e +e − Collisions, JETP Lett. 46 (1987) 525

47. A.H. Hoang et al., Top-antitop pair production close to threshold: Synopsis of recent NNLO results, Eur. Phys. J. direct C 3 (2000) 1

48. M. Beneke, A. Signer and V.A. Smirnov, Top quark production near threshold and the top quark mass, Phys. Lett. B 454 (1999) 137

49. A. Pineda and A. Signer, Heavy Quark Pair Production near Threshold with Potential Non-Relativistic QCD, Nucl. Phys. B 762 (2007) 67

50. M. Beneke, P. Falgari and C. Schwinn, Threshold resummation for pair production of coloured heavy (s)particles at hadron colliders, Nucl. Phys. B 842 (2011) 414

51. S. Catani, M.L. Mangano, P. Nason and L. Trentadue, The Resummation of soft gluons in hadronic collisions, Nucl. Phys. B 478 (1996) 273

52. W. Beenakker, S. Brensing, M. Kr¨amer, A. Kulesza, E. Laenen and I. Niessen, NNLL resummation for squark-antisquark pair production at the LHC, JHEP 01 (2012) 076

53. W. Beenakker et al., Towards NNLL resummation: hard matching coefficients for squark and gluino hadroproduction, JHEP 10 (2013) 120

54. M. Czakon and A. Mitov, On the Soft-Gluon Resummation in Top Quark Pair Production at Hadron Colliders, Phys. Lett. B 680 (2009) 154

55. L.A. Harland-Lang, A.D. Martin, P. Motylinski and R.S. Thorne, Parton distributions in the LHC era: MMHT 2014 PDFs, Eur. Phys. J. C 75 (2015) 204

56. M. Bonvini et al., Parton distributions with threshold resummation, JHEP 09 (2015) 191

57. S. Catani, D. de Florian, M. Grazzini and P. Nason, Soft gluon resummation for Higgs boson production at hadron colliders, JHEP 07 (2003) 028

58. A. Broggio, A. Ferroglia, B.D. Pecjak, A. Signer and L.L. Yang, Associated production of a top pair and a Higgs boson beyond NLO, arXiv:1510.01914

CITATION

CHAPTER 1

Brian Robson (2012). The Generation Model of Particle Physics, Particle Physics, Dr. Eugene Kennedy (Ed.), ISBN: 978-953-51-0481-0.

CHAPTER 2

O. Zeynali, D. Masti, M. Nezafat and A. Mallahzadeh, "Study of "Radiation Effects of Nuclear High Energy Particles" on Electronic Circuits and Methods to Reduce Its Destructive Effects," *Journal of Modern Physics*, Vol. 2 No. 12, 2011, pp. 1567-1573. doi: 10.4236/jmp.2011.212191.

CHAPTER 3

El-Bakry, M. , El-Dahshan, E. , Radi, A. , Tantawy, M. and Moussa, M. (2016) Modeling and Simulation for High Energy Sub-Nuclear Interactions Using Evolutionary Computation Technique. *Journal of Applied Mathematics and Physics*, 4, 53-65. doi: 10.4236/jamp.2016.41009.

CHAPTER 4

Haque, M. , Tariq, M. and Hussain, T. (2014) Presence of Multifractality in High-Energy Nuclear Collisions. *Journal of Modern Physics*, 5, 1889-1895. doi: 10.4236/jmp.2014.517183.

CHAPTER 5

Teresa Marrodán Undagoitia and Ludwig Rauch, Dark matter direct-detection experiments, Doi:10.1088/0954-3899/43/1/013001.

CHAPTER 6

Paola La Rocca and Francesco Riggi (2011). The Use of Avalanche Photodiodes in High Energy Electromagnetic Calorimetry, Advances in Photodiodes, Prof. Gian Franco Dalla Betta (Ed.), ISBN: 978-953- 307-163-3.

CHAPTER 7

Kihyeon Cho (2012). The e-Science Paradigm for Particle Physics, Particle Physics, Dr. Eugene Kennedy (Ed.), ISBN: 978-953-51-0481-0.

CHAPTER 8

Joseph John Bevelacqua (2012). Muon Colliders and Neutrino Effective Doses, Particle Physics, Dr. Eugene Kennedy (Ed.), ISBN: 978-953-51-0481-0.

CHAPTER 9

Anna Kulesza, Leszek Motyk, Tomasz Stebel and Vincent Theeuwes, Soft gluon resummation for associated $t\bar{t}H$ production at the LHC, Doi: 10.1007/JHEP03(2016)065.

INDEX